Core Books in Advanced Mathematics

Curve sketching

Core Books in Advanced Mathematics

General Editor: C. PLUMPTON, Moderator in Mathematics,
University of London School Examinations Department;
formerly Reader in Engineering Mathematics,
Queen Mary College, University of London.

Advisory Editor: N. WARWICK

Titles available

Differentiation
Integration
Vectors
Curve sketching
Newton's Laws and Particle Motion
Mechanics of Groups of Particles
Methods of Trigonometry
Coordinate Geometry and Complex Numbers
Proof
Methods of Algebra
Statistics
Probability

Core Books in Advanced Mathematics

Curve sketching

H. M. Kenwood
Chief Examiner A/O Level Mathematics,
University of London School Examinations
Department; Director of Studies, King
Edward's School, Bath.

C. Plumpton
Moderator in Mathematics, University of
London School Examinations Department;
formerly Reader in Engineering Mathematics,
Queen Mary College, University of London.

MACMILLAN
EDUCATION

First published 1983
Reprinted 1986

Published by
MACMILLAN EDUCATION LTD
Houndmills, Basingstoke, Hampshire RG21 2XS
and London
Companies and representatives
throughout the world

Printed in Hong Kong

British Library Cataloguing in Publication Data
Kenwood, H. M.
Curve sketching. – (Core books in advanced
mathematics)
1. Curve sketching
I. Title   II. Plumpton, C.
III. Series
519.5'32   QA297.6
ISBN 0–333–34803–6

Contents

Preface

Advanced level mathematics syllabuses are once again undergoing changes of content and approach, following the revolution in the early 1960s which led to the unfortunate dichotomy between 'modern' and 'traditional' mathematics. The current trend in syllabuses for Advanced level mathematics now being developed and published by many GCE Boards is towards an integrated approach, taking the best of the topics and approaches of the modern and traditional, in an attempt to create a realistic examination target through syllabuses which are maximal for examining and minimal for teaching. In addition, resulting from a number of initiatives, core syllabuses are being developed for Advanced level mathematics syllabuses, consisting of techniques of pure mathematics as taught in schools and colleges at this level.

The concept of a core can be used in several ways, one of which is mentioned above namely the idea of a core syllabus to which options such as theoretical mechanics, further pure mathematics and statistics can be added. The books in this series are core books involving a different use of the core idea. They are books on a range of topics, each of which is central to the study of Advanced level mathematics; they form small core studies of their own, of topics which together cover the main areas of any single-subject mathematics syllabus at Advanced level.

Particularly at times when economic conditions make the problems of acquiring comprehensive textbooks giving complete syllabus coverage acute, schools and colleges and individual students can collect as many of the core books as they need, one or more, to supplement books already possessed, so that the most recent syllabuses of, for example, the London, Cambridge, AEB and JMB GCE Boards, can be covered at minimum expense. Alternatively, of course, the whole set of core books gives complete syllabus coverage of single-subject Advanced level mathematics syllabuses.

The aim of each book is to develop a major topic of the single-subject syllabuses giving essential book work and worked examples and exercises arising from the authors' vast experience of examining at this level and including actual past GCE questions also. Thus, as well as using the core books in either of the above ways, they would also be ideal for supplementing comprehensive textbooks in the sense of providing more examples and exercises, so necessary for preparation and revision for examinations on the Advanced level mathematics syllabuses offered by the GCE Boards.

The ability to sketch a curve, given in cartesian, parametric or polar form, is essential for A level mathematics and is the key to success in many aspects of the subject. In this particular book, we cover, from first principles, the requirements of non-specialist mathematicians in accordance with the core syllabus of pure mathematics now being included by GCE Examining Boards in Advanced level syllabuses and meet the requirements of the polytechnics and universities for entrants to degree courses in mathematics-related subjects.

The importance of technique – i.e. choice of suitable method – in tackling problems has been stressed, with the statement and proof of standard results kept to a minimum, or covered by worked examples. While inevitably lacking experience, the student should try to acquire and appreciate good technique, so that more difficult problems can be tackled confidently. Only the most elementary knowledge of coordinates has been assumed, and important basic results are listed for easy reference.

Plenty of examples are provided throughout the book, both as exercises and as part of the text; the worked examples sometimes give a comparison between good and bad methods. We are grateful to the following GCE Examining Boards for permission to reproduce questions from past Advanced level GCE papers:

University of London Entrance and School Examinations Council (L)
The Oxford Delegacy of Local Examinations (O)
University of Cambridge Local Examinations Syndicate (C)
Oxford and Cambridge Schools Examination Board (OC)
The Northern Universities Joint Matriculation Board (JMB)
The Associated Examining Board (AEB)
The Welsh Joint Education Committee (W)

We are also grateful to Mrs Elizabeth Gilbert who typed the manuscript so well.

H. M. Kenwood
C. Plumpton

1   Algebraic curves

1.1   Introduction

Much work in mathematical analysis is centred around the concept of a function. Expressions like $x^3 - 3x + 1$ have little significance until some values are specified for x. When values are specified, the expression $x^3 - 3x + 1$ is said to be a function of x and we write

$$\text{f}: x \mapsto x^3 - 3x + 1 = y, \quad \text{say.}$$

If $x = 3$, we can write either $y = 3^3 - 3^2 + 1 = 19$ or $\text{f}(3) = 19$, and, if $x = \frac{1}{2}$, we write $y = (\frac{1}{2})^3 - 3/2 + 1 = -3/8$, or $\text{f}(\frac{1}{2}) = -3/8$. Values of x, called the *independent variable*, may be arbitrarily chosen from a well-defined set which is called the *domain* of the function. Each member x of the domain (set) will produce from the formula defining the function under consideration (sometimes called a mapping) one element y, called the *dependent variable*; all such elements y form a second set called the *range* (set) of the function.

In this book we shall be considering the graphs of functions whose domains belong to either the set of real numbers $\mathbb{R}$ or continuous intervals within the set, for example $a \leqslant x \leqslant b$. The behaviour of a function is often most clearly shown by a sketch-graph. Such a sketch should indicate the general shape of the graph over the whole domain of the function. This sketch may show particular important points, such as intersections with the coordinate axes, but, in general, you should not need to construct a detailed table of values which would be vital if you were actually making an accurate drawing of the graph.

The existence of a unique inverse function f^{-1} of a function f, given by $\text{f}: x \mapsto \text{f}(x) = y$ where $x \in \mathbb{R}$, can be seen at once by inspection of the graph of f. The inverse function f^{-1} will exist only if to each value of y there corresponds one and only one value of x; that is, the function f is necessarily a $1:1$ mapping. In terms of the graph of f this means that no line parallel to the x-axis intersects this graph in more than one distinct point. If this requirement is fulfilled, then the graph of f^{-1} can be obtained at once by reflecting the graph of f in the line $y = x$. It is often necessary to restrict the domain of a function to a subset of $\mathbb{R}$ in order to satisfy the $1:1$ mapping condition, which will then imply the existence of an inverse function. The inverse trigonometric functions discussed in §2.2 (p. 30) illustrate this requirement.

1.2 The straight line and line-pairs

The general equation of a straight line l is

$$ax + by + c = 0,$$

where a, b, c are constants.

The intersection of l with the x-axis is $A(-c/a, 0)$.

The intersection of l with the y-axis is $B(0, -c/b)$.

The gradient of l is $-a/b$.

How to sketch the line l

Draw the line passing through the points A and B.

Use the coordinates of a third point found from the equation of l to check the line you have sketched.

Special cases

Any line parallel to Ox is of the form $y = $ constant.

Any line parallel to Oy is of the form $x = $ constant.

Any line through O is of the form $y = mx$, where m is the gradient of the line.

Different forms of the linear equation

The line $y = mx + c$ has gradient m and passes through the point $(0, c)$.

The line $y - k = m(x - h)$ has gradient m and passes through the point (h, k).

The line $x/a + y/b = 1$ intercepts the coordinate axes at the points $(a, 0)$ and $(0, b)$.

The line $x \cos \alpha + y \sin \alpha = p$, where $p > 0$ and $0 \leqslant \alpha < 2\pi$; p is the length of the perpendicular OP from the origin to the line and α is the angle xOP, measured from Ox to OP with the usual sign convention used in trigonometry.

Some properties of lines

Two distinct lines have gradients m_1 and m_2.

The acute angle between the lines is $\tan^{-1} \left| \dfrac{m_1 - m_2}{1 + m_1 m_2} \right|$.

If $m_1 = m_2$, the lines are parallel.

If $m_1 m_2 + 1 = 0$, the lines are perpendicular.

Line-pairs through the origin

The equations of the lines $y - m_1 x = 0$ and $y - m_2 x = 0$ may be combined and written as

$$(y - m_1 x)(y - m_2 x) = 0,$$

called a line-pair.

A line-pair through the origin is represented by a second-degree equation, homogeneous in x and y.

If the line-pair given by $ax^2 + 2hxy + by^2 = 0$ represents the lines

$$(y - m_1x)(y - m_2x) = 0,$$

$$y^2 - (m_1 + m_2)xy + m_1m_2x^2 \equiv y^2 + 2hxy/b + ax^2/b = 0$$

and, in particular,

$$m_1 + m_2 = -2h/b \quad \text{and} \quad m_1m_2 = a/b.$$

Example 1   In one diagram sketch the lines with equations

$$x - 3 = 0, \quad 4x - 3y = 0, \quad 2x - 3y - 6 = 0, \quad 2x + 3y - 18 = 0.$$

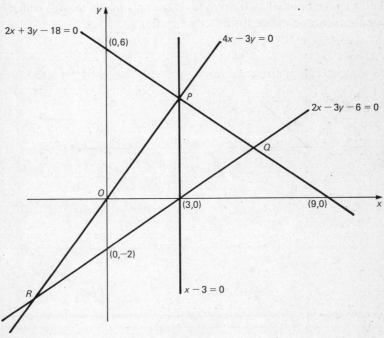

Fig. 1.1

Figure 1.1 shows the four lines.

Working exercise

By solving appropriate pairs of equations simultaneously, prove that the coordinates of the points P, Q and R as marked in Figure 1.1 are $(3, 4)$, $(6, 2)$ and $(-3, -4)$, respectively.

Prove by calculation that angle $PRQ = \tan^{-1}(6/17)$.

Example 2   The equation
$$2x^2 + 3xy - 2y^2 = 0 \equiv (2x - y)(x + 2y) = 0$$
represents the perpendicular straight lines $2x - y = 0$, $x + 2y = 0$.

Example 3   Draw, on separate diagrams, for $x \in \mathbb{R}$, graphs of the functions
(a) f: $x \mapsto x$, for $x \geqslant 0$,
   f: $x \mapsto -x$, for $x < 0$,
[The equation of this graph is written as $y = |x|$.]
(b) f: $x \mapsto [x]$, where $[x]$ represents the greatest integer not exceeding x,
(c) f: $x \mapsto |x + 2| + |x - 1|$,
(d) f: $x \mapsto x + 2$, for $x \geqslant 0$,
   f: $x \mapsto x - 1$, for $x < 0$.
Explain why the function defined in (d) is the only one of (a)–(d) which has an inverse function and define this inverse function.

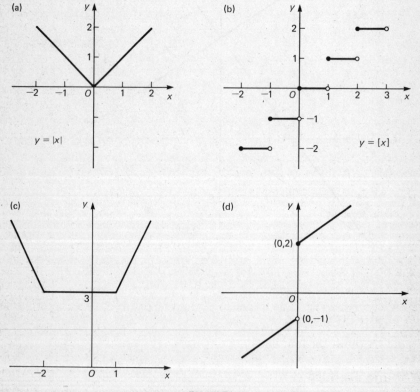

Fig. 1.2

The sketches shown in Fig. 1.2 represent the four functions.

In (a), (b) and (c) the function has no inverse function because, in each case, we see that different elements in the domain set of the function can produce the same element in the range set and this implies that none of the mappings is $1:1$. In (d) the mapping is $1:1$ and the inverse function f^{-1} is defined as

$$f^{-1}: x \mapsto x - 2, \quad \text{if } x \geqslant 2,$$
$$f^{-1}: x \mapsto x + 1, \quad \text{if } x < -1.$$

Working exercise

Copy the sketch of the graph of the function f defined in Example 3 (d). By reflecting your sketch in the line $y = x$, show that the graph of the inverse function f^{-1} is obtained and agrees with the definition of f^{-1} given above.

1.3   Regions defined by linear inequalities

The plane of the coordinate axes Oxy is divided into two regions by the line given by

$$f(x, y) \equiv ax + by + c = 0.$$

All points in the plane for which $f(x, y) = 0$ lie on the line.

For all other points in the plane, either $f(x, y) > 0$ or $f(x, y) < 0$. Further, $f(x, y)$ cannot change sign from positive at the point P, say, to negative at the point Q, say, or vice versa, without passing through zero. Hence, the line joining P and Q must cross the given line, placing P in one region and Q in the other. It follows that all points in one region into which the plane is divided by the line will have either the property $f(x, y) > 0$ or the property $f(x, y) < 0$. To find the sign of $f(x, y)$ in either region, it is sufficient to consider the sign at some conveniently chosen point, often merely the origin itself.

Example 4   The line given by

$$f(x, y) \equiv 3x + 4y - 12 = 0$$

divides the plane of the coordinate axes into two regions. Since $f(0, 0) = -12 < 0$, all points in the region which contains the origin O will have the property $f(x, y) < 0$ and all points in the region not containing O will have the property $f(x, y) > 0$.

Example 5   In Fig. 1.1, the inequalities $2x + 3y - 18 \leqslant 0, 4x - 3y \geqslant 0$ and $2x - 3y - 6 \leqslant 0$ are simultaneously satisfied by all points (x, y) which lie within or on the perimeter of triangle PQR.

The inequalities $2x + 3y - 18 < 0, 4x - 3y > 0$ and $2x - 3y - 6 < 0$ are simultaneously satisfied by all points (x, y) which lie in the triangle PQR but not including the perimeter.

Exercise 1.3

1 Mark the points $A(-1, -2)$, $B(2, 1)$, $C(1, 4)$ and $D(-2, 1)$ on a sketch and join the lines AB, BC, CD and DA. Prove that the quadrilateral $ABCD$ is a parallelogram.

2 Sketch the locus of points in the plane of the coordinate axes which are equidistant from the fixed points $A(-3, 1)$ and $B(5, -3)$. Find the equation of the locus.

3 Show in a sketch the triangular region S defined by the inequalities

$$3x - y + 4 > 0, \quad 24x + 7y - 3 < 0, \quad 3x - 4y - 15 < 0.$$

Verify that the vertices of S form an isosceles triangle.

4 Sketch the line-pair given by $2y^2 + xy - 6x^2 = 0$.
(a) Calculate the acute angle between the members of the line-pair.
(b) The line-pair meets the line $2x + 3y - 12 = 0$ at the points A and B. Calculate the area of $\triangle OAB$, where O is the origin.

5 Prove that the lines represented by the equation $3y^2 - 8xy - 3x^2 = 0$ are perpendicular. Two adjacent vertices O, A of a square $OABC$ have coordinates $(0, 0)$, $(1, 3)$, respectively. Show that the other two vertices, B and C, can occupy two possible positions and determine their coordinates.

6 Leave unshaded in a sketch the region T satisfied by the inequalities $2x + 3y - 6 > 0$, $4y - x - 8 < 0$ and $2y - x - 2 > 0$. If x and y were restricted to integral values only, find all the possible solution sets of these inequalities.

7 For $x \in \mathbb{R}$, the function f is defined by

$$\text{f}: x \mapsto \begin{cases} -2 + x/3 & \text{for } x \leqslant -3, \\ x & \text{for } -3 < x < 3, \\ 2 + x/3 & \text{for } x \geqslant 3. \end{cases}$$

(a) Sketch the graph of the function f.
(b) Sketch, on the same diagram, the graph of the inverse function f^{-1}.

8 Sketch, on the same diagram, the graphs of the functions given by
(a) $y = |x|$,   (b) $y = |3 - x|$,
(c) $y = |x| + |3 - x|$, (d) $y = |x| - |3 - x|$.

9 For $x \in \mathbb{R}$, the functions f and g are defined by

$$\text{f}: x \mapsto 3x - 2, \quad \text{g}: x \mapsto 2 - x.$$

(a) Sketch the graphs of f, g, fg and f^{-1}.
(b) Prove that $\text{g} = \text{g}^{-1}$ and that $\text{fg} = \text{gf}$.

10 Clearly indicating both included and excluded points, sketch graphs of the functions given by (a) $y = [x] - x$, (b) $y = x - [x]$.

1.4 The quadratic function $\text{f}: x \mapsto ax^2 + bx + c$, where $x \in \mathbb{R}$, $a \neq 0$

$$y = ax^2 + bx + c = a[x^2 + bx/a + c/a]$$
$$= a[(x + b/2a)^2 - (b^2 - 4ac)/(4a^2)].$$

Figure 1.3 illustrates the possible cases which can arise for the graph of f. In each case, the curve is a parabola.

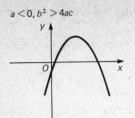

a < 0, b² > 4ac

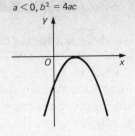

a < 0, b² = 4ac

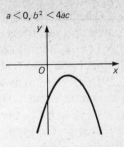

a < 0, b² < 4ac

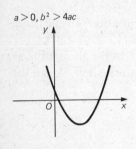

a > 0, b² > 4ac

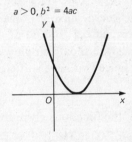

a > 0, b² = 4ac

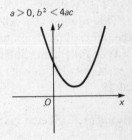

a > 0, b² < 4ac

Fig. 1.3

When $a < 0$, the parabola will be of shape ∩, because the x^2 term (the one of highest power) will be negative and will dominate the remaining terms when $|x|$ is large.

When $a > 0$, the parabola will be of shape ∪.

The solutions of the equation $ax^2 + bx + c = 0$ are given by the points at which the parabola $y = ax^2 + bx + c$ meets the x-axis ($y = 0$), if any such points exist.

The solutions are $[-b + \sqrt{(b^2 - 4ac)}]/(2a)$, $[-b - \sqrt{(b^2 - 4ac)}]/(2a)$.

The line $x + b/(2a) = 0$ is an axis of symmetry for the parabola $y = ax^2 + bx + c$.

The stationary point on this curve is $[-b/(2a), (4ac - b^2)/(4a)]$.

1.5 The cubic function f: $x \mapsto ax^3 + bx^2 + cx + d$, where $x \in \mathbb{R}$

Taking $a > 0$, the graph of f has the following properties.

(i) As $x \to +\infty$, $y \to +\infty$ because the x^3 term is positive and will dominate the remaining terms when x is large.

(ii) As $x \to -\infty$, $y \to -\infty$.

(iii) A consideration of (i) and (ii) implies that the graph of f must cross the x-axis at least once; taking this point as $x = \beta$, say, we have

$$ax^3 + bx^2 + cx + d \equiv (x - \beta)Q,$$

where Q is a quadratic expression in x.

(iv) The equation $Q = 0$ may have two real distinct roots, two real coincident roots or no real roots.

(v) $f'(x) = 3ax^2 + 2bx + c$ and the equation $f'(x) = 0$ may have two real distinct roots, two real coincident roots or no real roots.

If $f'(x) = 0$ has two real distinct roots, p and q say, where $p > q$, $f(p)$ is a minimum value of $f(x)$ and $f(q)$ is a maximum value of $f(x)$.

If $f'(x) = 0$ has two real coincident roots, r say, then the stationary value of $f(x)$ at $[r, f(r)]$ is a point of inflexion.

If $f'(x) = 0$ has no real roots, the graph of f has no stationary points.

(vi) $f''(x) = 6ax + 2b$ and $f''(x) = 0$, $f'''(x) \neq 0$ when $x = -b/(3a)$.

This implies that all cubic curves have a point of inflexion. In our case, the point of inflexion has coordinates $[-b/(3a), f(-b/3a)]$.

When sketching cubic curves you should employ as few of the above properties as possible in each case. Once you have sufficient information, sketch the curve.

Example 5 Sketch graphs of the functions given by
(a) $y = x^3 - 3x^2 + 2x$ $\qquad \equiv y = x(x - 1)(x - 2)$,
(b) $y = x^2 - x^3$ $\qquad\qquad \equiv y = x^2(1 - x)$,
(c) $y = x^3 - x^2 + x - 1$ $\quad \equiv y = (x^2 + 1)(x - 1)$,
(d) $y = -x^3 + 2x^2 - x + 2 \equiv y = (x^2 + 1)(2 - x)$.

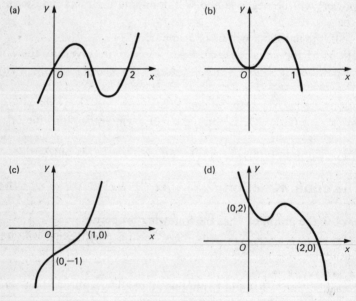

Fig. 1.4

The general shapes of the four graphs are shown in Fig. 1.4.

The graphs for (a) and (b) can be drawn at once because in (a) there are 3 intersection points with the x-axis and in (b) the double intersection at $x = 0$ implies that the x-axis is a tangent to the curve at the origin. In (c) there is only one real intersection point and the equation $\dfrac{dy}{dx} = 3x^2 - 2x + 1 = 0$ has no real roots $\Rightarrow$ there are no stationary points. In (d) there is only one real intersection point but the equation

$$\frac{dy}{dx} = -3x^2 + 4x - 1 = -(x - 1)(3x - 1) = 0$$

$\Rightarrow$ a maximum at $(1, 2)$ and a minimum at $(1/3, 50/27)$.
A check on the point of inflexion for each curve is left as an exercise for the reader.

Exercise 1.5

Assume that $x \in \mathbb{R}$ in this exercise.

1. Sketch, on separate diagrams, the graphs of the functions given by
   (a) $y = x^2 + 4x + 4$, (b) $y = x^2 - 4x - 4$, (c) $y = x^2 + 4$,
   (d) $y = 4 - x^2$, (e) $y = -4 + x - x^2$.
   In each case, find (i) the equation of the axis of symmetry, (ii) the range of the function.

2. Sketch the graphs of the functions given by
   (a) $y = (x - 1)(x - 4)$, (b) $y = |(x - 1)(x - 4)|$.
   Note: All that is needed for (b) is to replace that part of graph (a) which is below the x-axis by its reflection in this axis and to leave the part of (a) which is above the x-axis unchanged.

3. Sketch, in a single diagram, the graphs of the functions given by
   (a) $y = x^2$, (b) $y = 2 - x^2$, (c) $y = -(2x)^2$,
   (d) $y = |2 - x^2|$, (e) $y = -\sqrt{x}$.

4. Sketch, on separate diagrams, the graphs of the functions given by
   (a) $y = x^3 - x$, (b) $y = x^3 + x$, (c) $y = x^2 - x^3$.

5. The cubic curve given by $y = px^3 + qx + r$ passes through the point $(0, 6)$ and the tangent to the curve at the point $(1 \cdot 5, 3)$ is parallel to the x-axis. Find the values of p, q and r and hence sketch the curve.

6. Given that $f(x) \equiv x^3 - 6x^2 + 12x - 7$, prove that
   $$f'(2) = f''(2) = 0.$$
   Sketch the graph of the curve whose equation is
   $$y = x^3 - 6x^2 + 12x - 7.$$

7. Extend the argument given in §1.3 to include a region in the plane of the coordinate axes defined by $y < ax^2 + bx + c$.

1.6 Some properties of functions

Even functions

A function f is said to be an even function of x if

$$f(x) = f(-x) \text{ for every value of } x.$$

Examples are given by $f(x) = x^2$, $f(x) = 1/(1 + x^4)$, and any function of x whose power series expansion contains only even powers of x, for example, $\cos x$.

The graph of an even function is symmetrical about the y-axis.

Odd functions

A function f is said to be an odd function of x if

$$f(x) = -f(-x) \text{ for every value of } x.$$

Examples are given by $f(x) = x$, $f(x) = 3x - x^3/2$, and any function of x whose power series expansion contains only odd powers of x, for example, $\sin x$.

The graph of an odd function has point symmetry about the origin.

Periodic functions

A function f is said to be periodic if

$$f(x) = f(x + nk), \quad \text{where } k \text{ is a constant and } n \in \mathbb{Z}.$$

The period of the function is k.
Examples are given by $f(x) = \sin x$ (period 2π), $f(x) = \cos 3x$ (period $2\pi/3$)   and $f(x) = x - [x]$ (period 1).

Continuous functions

The function f whose domain is D is said to be continuous at $x = c$, where $c \in D$, if

$$\lim_{x \to c-} f(x) = f(c) = \lim_{x \to c+} f(x).$$

Intuitively you may take a function to be continuous throughout its domain if it is possible to sketch the whole graph without lifting your pen from the paper. Any break or jump in your graph will correspond to a discontinuity. See Fig. 1.2(d) (p. 4) and Fig. 1.6 (p. 12) for illustrations of discontinuities at $x = 0$.

Example 6   The function f has the following properties:
  (i) $f(x) = 3x$ for $0 \leqslant x < 1$,
 (ii) $f(x) = 4 - x$ for $1 \leqslant x \leqslant 4$,
(iii) $f(x)$ is an odd function,
(iv) $f(x)$ is periodic with period 8.
Sketch the graph of $f(x)$ for $-4 \leqslant x \leqslant 12$.

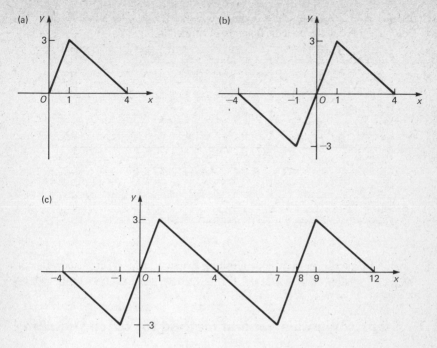

Fig. 1.5

For $0 \leqslant x \leqslant 4$ the graph of f consists of the line $y = 3x$ from $(0, 0)$ to $(1, 3)$ and then the line $x + y = 4$ from $(1, 3)$ to $(4, 0)$ as shown in Fig. 1.5 (a). Since f is an odd function, its graph has point symmetry about the origin and the graph of f for $-4 \leqslant x \leqslant 4$ is as shown in Fig. 1.5 (b). Finally, since f is periodic with period 8, $f(x + 8) = f(x)$ for all x; the graph of f for $4 \leqslant x \leqslant 12$ is obtained by translation through 8 parallel to Ox of the part of the graph between $(-4, 0)$ and $(4, 0)$.

The whole graph required is shown in Fig. 1.5 (c).

Exercise 1.6

1 The function f is periodic with period 3 and

$$f(x) = \sqrt{(25 - 4x^2)} \quad \text{for } 0 < x \leqslant 2,$$
$$f(x) = 9 - 3x \quad \text{for } 2 < x \leqslant 3.$$

Sketch the graph of f for $-3 \leqslant x \leqslant 6$.

2 For $x \geqslant 0$ a function f is given by $f(x) = x^3 - 4x$.
Sketch the complete graph of f when   (a) f is an odd function,   (b) f is an even function.

3 Determine, for each of the following functions, whether it is odd, even, periodic or none of these.
(a) $f(x) = x^2(1 - x^2)$ for $x \in \mathbb{R}$,
(b) $f(x) = 3x(1 + x^2)$ for $x \in \mathbb{R}$,
(c) $f(x) = x^2(1 - x)$ for $x \in \mathbb{R}$,

(d) $f(x) = x^2 + 1$ for $0 < x \leqslant 3$ and also $f(x) = f(x + 3)$ for all values of x.

4 Completely define the periodic function illustrated in Fig. 1.6.

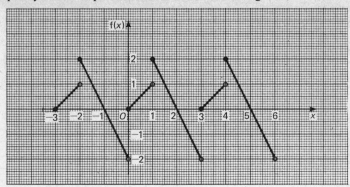

Fig. 1.6

5 Given that $g(x)$ is an even non-vanishing function and $h(x)$ is an odd function, determine whether (a) $h(x)/g(x)$, (b) $hg(x)$, (c) $gh(x)$ are necessarily odd, even or neither.

1.7 Steps towards a general method for curve sketching

Implicit relations between two variables

Frequently relations between two variables x and y are given implicitly by an algebraic equation in the form $f(x, y) = 0$. If x is taken as the independent variable, more than one value of y may result for a value of x, and vice versa. In such cases the word *function* should be avoided, because we have asserted that y is a function of x when, for each value of x, there exists one and only one corresponding value of y. Some simple and very well-known curves fall into this category and we shall simply call them curves and state their equations.

Example 7 Consider the curve $x^2 + y^2 - 4 = 0$.
For any point $P(x, y)$ in the plane of the coordinate axes Oxy, $OP^2 = x^2 + y^2$ (Pythagoras). The point P will lie on the curve $x^2 + y^2 - 4 = 0$ if and only if $OP = 2$. This curve is a circle, centre O and radius 2, and we can sketch the curve at once.
For $|x| < 2$, two values of y are obtained for each value of x.
For $|x| = 2$, $y = 0$ and for $|x| > 2$, no real values of y exist.
A sketch of the circle is shown in Fig. 1.7 (a).

Curves with asymptotes parallel to the coordinate axes

Example 8 The curve $y = 1/x$ has no value of y corresponding to $x = 0$. If $x > 0$ and $x \to 0$, then y is positive and increases without limit and we write $y \to +\infty$.

Similarly, if $x < 0$ and $x \to 0$, then $y \to -\infty$. The curve $y = 1/x$ is said to approach the line $x = 0$ asymptotically as this line recedes from the origin.

The equation of the curve may also be written as $x = 1/y$ and the curve will also approach the line $y = 0$ asymptotically as this line recedes from the origin.

The lines $x = 0$ and $y = 0$ are called *asymptotes* and a sketch of the curve, called a rectangular hyperbola, is shown in Fig. 1.7(b).

Note: It is essential when sketching a curve which has discontinuities to determine on which side of each asymptote the curve lies.

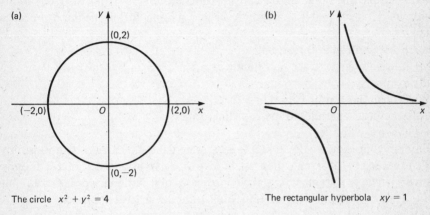

(a)

(0,2)

(−2,0)   O   (2,0) x

(0,−2)

The circle $x^2 + y^2 = 4$

(b)

O   x

The rectangular hyperbola $xy = 1$

Fig. 1.7

Symmetrical properties of curves

The following symmetrical properties refer to the curve $f(x, y) = 0$.

(i) If $f(-x, y) = 0$, the curve is symmetrical about Oy.
(ii) If $f(x, -y) = 0$, the curve is symmetrical about Ox.
(iii) If $f(y, x) = 0$, the curve is symmetrical about the line $y - x = 0$.
(iv) If $f(-y, -x) = 0$, the curve is symmetrical about the line $y + x = 0$.
(v) If $f(-x, -y) = 0$, the curve has point symmetry about O.

The degree of a curve

When the equation of an algebraic curve is given implicitly in the form $f(x, y) = 0$, where $f(x, y)$ can be expressed as a polynomial in x and y, we define the degree of the curve to be the highest combined power of x and y occurring in any term.

Example 9 State the degree of each of the following curves:
(a) $xy = x + 1$,
(b) $(3y - x^2)(x - 3y^2) = 0$,
(c) $y = x/(x^4 + 3)$.

(a) The degree of the curve is 2 on account of the term xy.

(b) The degree of the curve is 4 on account of the term $3x^2y^2$.

(c) $y = x/(x^4 + 3) \Leftrightarrow (x^4 + 3)y = x$ and the degree of the curve is 5 on account of the term x^4y.

An important check on a sketch

Any straight line cuts a curve of degree n in n points *at most*.

Summary

Many simple algebraic curves can be sketched directly from their equations without detailed analysis. Examine each equation for particular properties such as

  (i) passage through the origin,

 (ii) intersections with the coordinate axes,

(iii) symmetry,

(iv) restrictions on the values of x or of y,

 (v) asymptotes parallel to the coordinate axes,

(vi) other asymptotes.

Build up your sketch as each fact is ascertained and use calculus methods only when necessary, although these methods can serve as a check or for finding some specific requirement, e.g. the coordinates of a stationary point or a point of inflexion. The following examples illustrate the general approach.

Example 10   Sketch the curve

$$y = \frac{(x + 6)(x - 1)}{(x - 3)(x + 2)}.$$

We first note the following:

  (i) when $y = 0$, $x = -6$ or $x = 1$,

 (ii) when $x = 0$, $y = 1$,

(iii) the lines $x - 3 = 0$ and $x + 2 = 0$ are asymptotes to the curve parallel to Oy,

(iv) For	$x < -6$	$-6 < x < -2$	$-2 < x < 1$	$1 < x < 3$	$x > 3$
Sign of y is	$+$	$-$	$+$	$-$	$+$

  (v) the equation of the curve may be written as $y = \dfrac{1 + 5/x - 6/x^2}{1 - 1/x - 6/x^2}$

$\Rightarrow$ as $x \to \infty$, $y \to 1 +$ (i.e. y approaches the value 1 from above),

and as $x \to -\infty$, $y \to 1 -$ (i.e. y approaches the value 1 from below),

$\Rightarrow$ the line $y - 1 = 0$ is an asymptote to the curve parallel to Ox.

These facts are shown in Fig. 1.8(a), where shading denotes regions in which the curve cannot lie, crosses show important plot points, and broken lines indicate the three asymptotes.

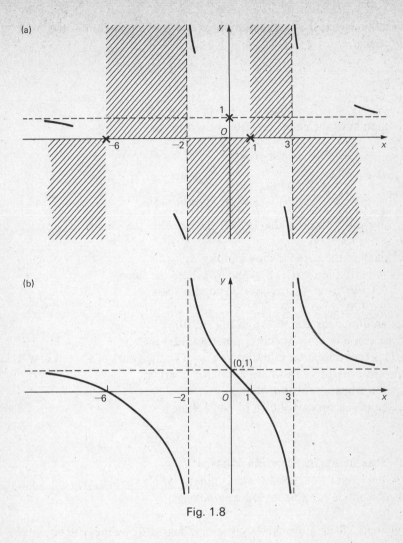

(a)

(b)

Fig. 1.8

The curve is sketched using Fig. 1.8(a) and is shown in Fig. 1.8(b).
Check: The degree of the curve is 3 on account of the term x^2y when the equation is written in the form $y(x^2 - x - 6) = x^2 + 5x - 6$, and we can see that a general straight line will cut the curve in 3 points at most, as is required.

Example 11 Given that $f(x) \equiv \dfrac{2x(2 - x)}{1 + x^2}$, sketch the curves

(a) $y = f(x)$, (b) $y = \dfrac{1}{f(x)}$, (c) $y^2 = f(x)$.

(a) Since for $x \in \mathbb{R}$, the expression $1 + x^2$ is never zero, the curve is continuous for all real x and there are no asymptotes parallel to Oy. Also the curve will cross the x-axis at the origin and when $x = 2$. The equation of the curve may be written as

$$y = \frac{4/x - 2}{1/x^2 + 1}.$$

As $x \to \infty$, $y \to -2+$ and as $x \to -\infty$, $y \to -2-$.
The equation of the curve may also be written as

$$(y + 2)x^2 - 4x + y = 0.$$

For real x,
$$16 \geqslant 4y(y + 2)$$

$$\Rightarrow y^2 + 2y - 4 \leqslant 0.$$

This restriction implies that y will lie between the values $(-\sqrt{5} - 1)$ and $(\sqrt{5} - 1)$ for all real x.
   A sketch of the curve is shown in Fig. 1.9(a).
(b) Note   (i) the signs of f(x) and $1/$f(x) are the same,
            (ii) as $y \to \infty$, $1/y \to 0$ and vice versa,
            (iii) for $y = \pm 1$, $1/y = \pm 1$.
A sketch of the curve is shown in Fig. 1.9(b).
(c) The graph of $y^2 =$ f(x) is symmetrical about Ox.
Also from consideration of the curve $y =$ f(x), the graph of $y^2 =$ f(x) will have
(i) no curve in intervals where $y < 0$ for $y =$ f(x),
(ii) $y = 0$ when $x = 0$ and when $x = 2$.
A sketch of the curve is shown in Fig. 1.9(c).

1.8   Parametric representation
It is often useful to express the coordinates of a variable point on a curve in terms of a single variable, called a *parameter*.

Example 12   Sketch the curves given parametrically by the equations
(a) $x = t^2$, $y = 2t$,
(b) $x = t^2$, $y = t^3$,
(c) $x = t^2$, $y = t(t^2 - 1)$.
In each case, give the cartesian equation of the curve.

Each of these curves passes through O, is symmetrical about Ox and no part of the curve lies to the left of the y-axis, since $x \geqslant 0$ for all t. The cartesian equation of each curve is obtained by eliminating t from the parametric equations.
(a) The cartesian equation is $y^2 = 4x$ and the curve, shown in Fig. 1.9(d), is a *parabola*. The y-axis is a tangent at the origin to the parabola.
(b) The cartesian equation is $y^2 = x^3$ and the curve, shown in Fig. 1.9(e), is called a *semi-cubical parabola*. The curve has two branches which meet at the

(a)
$y = f(x)$

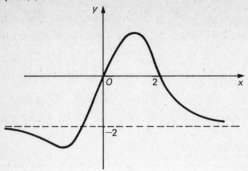

(d)

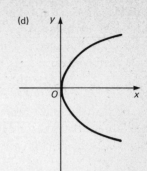

(b)
$y = 1/f(x)$

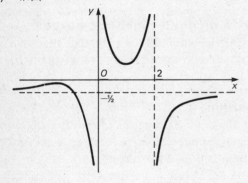

(e)

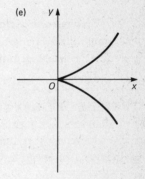

(c)
$y^2 = f(x)$

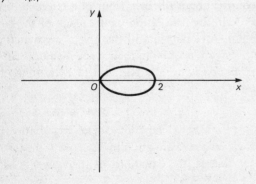

(f)

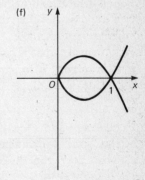

Fig. 1.9

origin and the x-axis is a tangent to each branch. The origin is a *double point* of the curve and is called a *cusp*.

(c) The cartesian equation is $y^2 = x(x-1)^2$ and the curve, shown in Fig. 1.9(f), is of degree 3. The curve has two branches which intersect at the point $(1, 0)$ where $t = \pm 1$.

For this curve, $\dfrac{dy}{dx} = \dfrac{dy}{dt} \bigg/ \dfrac{dx}{dt} = (3t^2 - 1)/(2t)$, and the gradients of the tangents to the two branches of the curve at $(1, 0)$ are 1 and -1 respectively; it follows that these tangents are perpendicular. The point $(1, 0)$ where the branches intersect is called a *node*.

Working exercise

Sketches of curves, some with intriguing names, are shown in Fig. 1.10. In each case, taking $a > 0$ and $b > 0$, investigate the properties of the curve from its equation and verify the general shape shown by producing a sketch of your own.

1.9 The sketching of more difficult algebraic curves

We conclude the chapter by giving a brief outline of the method of attack which is adopted when sketching algebraic curves with more complicated equations.

The form of a curve at the origin

For a curve which passes through the origin O, tangents to each branch of the curve which pass through O are obtained by equating to zero any linear factors of the terms of *lowest* degree in the equation of the curve.

Example 13 For the curve $y^2(a - x) = x^2(a + x)$, where $a > 0$ (see Fig. 1.10(e)) the equations of the tangents to the curve at O are given by $y^2a - x^2a = 0$, the terms of the lowest degree in the equation; the tangents are $y = x$ and $y = -x$.

Similar information can be acquired about the behaviour of a curve at any specific point A, other than the origin, by translation of the plane of the co-ordinate axes to make A the origin. Any tangents to the given curve at the origin provide what is called a *first approximation* to their respective branches near the origin; a *second approximation* is usually required to determine on which side of the tangent the branch lies.

Example 14 For the curve $x^3 + y^3 = 3axy$, where $a > 0$ (see Fig. 1.10(f)), the equations of the tangents to the curve at O are given by $3axy = 0$, the term of lowest degree in this equation: that is, the equations of the tangents are $x = 0$ and $y = 0$.

Name of curve	Cartesian equation	Parametric equations
(a) Ellipse	$b^2x^2 + a^2y^2 = a^2b^2$	$x = \dfrac{a(1 - t^2)}{1 + t^2}, y = \dfrac{2bt}{1 + t^2}$
(b) Hyperbola	$b^2x^2 - a^2y^2 = a^2b^2$	$x = \dfrac{a(1 + t^2)}{1 - t^2}, y = \dfrac{2bt}{1 - t^2}$
(c) Witch of Agnesi	$x(a^2 + y^2) = a^3$	$x = \dfrac{a}{1 + t^2}, \quad y = at$
(d) Cissoid of Diocles	$y^2(a - x) = x^3$	$x = \dfrac{at^2}{1 + t^2}, \quad y = \dfrac{at^3}{1 + t^2}$
(e) Strophoid	$y^2(a - x) = x^2(a + x)$	$x = \dfrac{a(t^2 - 1)}{t^2 + 1}, y = \dfrac{at(t^2 - 1)}{t^2 + 1}$
(f) Folium of Descartes	$x^3 + y^3 = 3axy$	$x = \dfrac{3at}{1 + t^3}, \quad y = \dfrac{3at^2}{1 + t^3}$

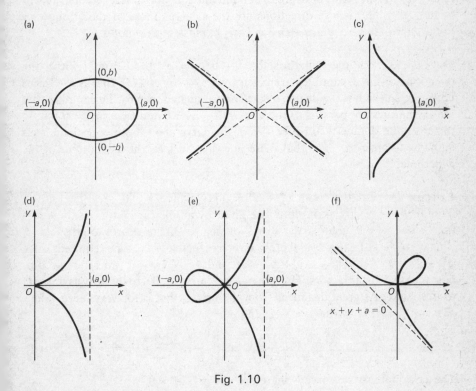

Fig. 1.10

Considering the branch of this curve for which $y = 0$ is a tangent at O, we take $x^3 - 3axy \approx 0$ as our second approximation, since near this tangent y is small and y^3 is very small. This gives the second approximation for this branch in the neighbourhood of O as $y \approx x^2/(3a)$.

Similarly, the other branch for which $x = 0$ is a tangent can be represented by the second approximation $x \approx y^2/(3a)$.

We see in Fig. 1.10(f) that each branch is approximately parabolic near O, thus agreeing with the second approximations.

Asymptotes

For any curve of degree n, it is usually possible to get some idea of the form of the curve at great distances from the origin by equating to zero those terms of degree n. If the terms of degree n contain linear factors, these factors equated to zero give lines whose directions are the same as those of the asymptotes to the curve. *A curve of degree n can have, at most, n asymptotes.*

Example 15 For the hyperbola $b^2x^2 - a^2y^2 = a^2b^2$, see Fig. 1.10(b), the directions of the asymptotes are given by $b^2x^2 - a^2y^2 = 0$, that is the terms of degree 2, which is the degree of the curve, equated to zero. This implies that the asymptotes are parallel to the lines $bx - ay = 0$ and $bx + ay = 0$ but, in this case, the symmetrical properties of the curve about both coordinate axes allow us to assert at once that these lines are, in fact, the asymptotes of the hyperbola.

Example 16 For the curve $x^3 + y^3 = 3axy$, where $a > 0$, see Fig. 1.10(f), the directions of any asymptotes are given by any linear factors in the equation $x^3 + y^3 = 0$, that is the terms of degree 3, the degree of the curve, equated to zero. In this case there is only one factor of the type required, since $x^3 + y^3 \equiv (x + y)(x^2 - xy + y^2) = 0$ and $x^2 - xy + y^2$ does not decompose into real linear factors. The line $x + y = 0$ is parallel to the only asymptote to this curve. At great distances from the origin, that is for very large values of x and y, we may write $y \approx -x$, and

$$x + y = \frac{3axy}{x^2 - xy + y^2} = \lim_{x \to \infty} \left[\frac{-3ax^2}{x^2 + x^2 + x^2} \right] = -a.$$

The asymptote to the curve is therefore $x + y + a = 0$.

A second approximation is obtained by putting $y = -x - a$ in the equation

$$x + y = \frac{3axy}{x^2 - xy + y^2}.$$

This gives

$$x + y \approx \frac{-3ax(x + a)}{x^2 + x(a + x) + (a + x)^2} = -a + \frac{a^3}{a^2 + 3ax + 3x^2}.$$

For all x, we note that $a^2 + 3ax + 3x^2 > 0$, and for large $|x|$, $x + y = -a + \varepsilon$, where ε is a small positive number which approaches zero as $|x| \to \infty$. As $x \to +\infty$ and as $x \to -\infty$, the curve lies on the same side of the asymptote as the origin.

Exercise 1.9

1   Given that n is a positive integer, sketch the following curves for (i) $a > 0$, (ii) $a < 0$.
   (a) $y = ax^{2n+1}$,   (b) $y = ax^{2n}$,   (c) $y = ax^{-1-2n}$,   (d) $y = ax^{-2n}$.

2   Sketch, on the same axes, the following pairs of curves:
   (a) $y = x^3$, $y^2 = x^3$,   (b) $y = x^2$, $y^2 = x^2$,
   (c) $y = x^3 - x^2$, $y^2 = x^3 - x^2$,   (d) $y = x^3 + x^2$, $y = 1/(x^3 + x^2)$.

3   Sketch, on separate axes, the following curves:
   (a) $y = x/(x^2 - 1)$,   (b) $y = x/(x^2 + 1)$,
   (c) $y = (x^2 + 1)/x$,   (d) $y^2 = x/(x^2 - 1)$.

4   Given that $x \in \mathbb{R}$, prove that

$$\frac{(2x - 1)(x - 8)}{(x - 1)(x - 4)}$$

   never takes a value between 25/9 and 9.
   Sketch the graphs of (a) $y = \dfrac{(2x - 1)(x - 8)}{(x - 1)(x - 4)}$,   (b) $y = \dfrac{(x - 1)(x - 4)}{(2x - 1)(x - 8)}$.

5   Prove that, if $x \in \mathbb{R}$, $\dfrac{x^2 - 4}{x^2 - 4x + 3}$ may take all real values.
   Sketch the graphs of (a) $y = \dfrac{x^2 - 4}{x^2 - 4x + 3}$,   (b) $y^2 = \dfrac{x^2 - 4}{x^2 - 4x + 3}$.

6   Given that $y = (x^2 - 1)/(x^2 + 1)$ and $x \in \mathbb{R}$, prove that $-1 \leqslant y < 1$.
   Sketch the graph of $y = (x^2 - 1)/(x^2 + 1)$.

7   State the equation of the tangent at the origin to the following curves:
   (a) $y(x^2 + 1) = x$,   (b) $y(x^2 + 1) = x^2$,   (c) $y(x^2 + 1) = x^3$.
   Use any symmetrical properties to sketch the curves.

8   Find the equations of the asymptotes to the curves
   (a) $x^2y = x^3 - 3x^2 + 4$,   (b) $y^2(x - y) = 2x$,   (c) $y^4 - x^4 = 4x^2y$.
   Sketch the curves.

9   Sketch the curves, represented parametrically by
   (a) $x = t - 1$, $y = t^2$,
   (b) $x = t/(1 + t^2)$, $y = (1 - t^2)/(1 + t^2)$,
   (c) $x = (t + 1)^2$, $y = t(t + 1)^2$,
   (d) $x = 2t/(1 + t^4)$, $y = 2t^3/(1 + t^4)$.
   In each case, find the cartesian equation of the curve.
   Note: (d) is called the *lemniscate of Bernoulli*.

10 Sketch the form of the *conchoid* given by the equation

$$y^2 = \frac{x^2(b - c + x)(b + c - x)}{(x - c)^2}, \quad \text{for } b, c > 0$$

when (a) $b < c$, (b) $b = c$, (c) $b > c$.

Miscellaneous Exercise 1

1 A curve C_1 has equation $2y^2 = x$; a curve C_2 is given by the parametric equations $x = 4t$, $y = 4/t$. Sketch C_1 and C_2 on the same diagram and calculate the coordinates of their point of intersection, P. **(L)**

2 The function f is defined by f: $x \mapsto \dfrac{x^2 + 3}{x}$ where $x \in \mathbb{R}$ and $x \neq 0$. Sketch the graph of f.

On the same diagram sketch the graph of g: $x \mapsto 2x$ where $x \in \mathbb{R}$. **(L)**

3 Shade on a sketch the domain for which

$$y^2 - x < 0 \quad \text{and} \quad x^2 - y < 0.$$ **(L)**

4 The function f is periodic with period 2 and

$$\begin{aligned} f(x) &= 4/x & \text{for } 1 < x \leqslant 2, \\ f(x) &= 2(x - 1) & \text{for } 2 < x \leqslant 3. \end{aligned}$$

Sketch the graph of f when $-3 \leqslant x \leqslant 3$. **(L)**

5 The function f has the following properties:
(a) $f(x) = 4x$ for $0 \leqslant x < 2/3$,
(b) $f(x) = 8(1 - x)$ for $2/3 \leqslant x \leqslant 1$,
(c) $f(x)$ is an odd function,
(d) $f(x)$ is periodic with period 2.
Sketch the graph of $f(x)$ for $-3 \leqslant x \leqslant 3$. **(L)**

6 Shade the region or regions of the x, y plane within which $(y^2 - 8x)(x^2 + y^2 - 9)$ is negative. **(L)**

7 Find the equation of each of the three asymptotes of the curve

$$y = \frac{2x^2}{3x^2 - x - 4}.$$ **(L)**

8 Find the set of values of x for which $|2x - 5| < |x|$. **(L)**

9 Show that the curves whose equations are

$$y - 1 = x^3 \quad \text{and} \quad y + 3 = 3x^2$$

intersect at a point on the x-axis, and find the coordinates of this point. Show also that the only other point at which the curves meet is $(2, 9)$, and that the curves have a common tangent there. Sketch the two curves on the same diagram. **(JMB)**

10 Find the coordinates of the points of intersection of the curves whose equations are

$$y = (x - 1)(x - 2), \quad y = \frac{3(x - 1)}{x}.$$

Sketch, or obtain:
(i) the coordinates of the turning point of the first curve;
(ii) the equations of the asymptotes of the second curve.
Sketch the two curves on the same diagram. **(JMB)**

11 Given that $f(x) \equiv x - 1 + \dfrac{1}{x+1}$, x real, $x \neq -1$, find the values of x for which $f'(x) = 0$. Sketch the graph of f, showing the coordinates of the turning points and indicating clearly the form of the graph when $|x|$ becomes large. (JMB)

12 Let $f(x) \equiv \dfrac{ax + b}{x + c}$ where x, a, b, c are real and $x \neq \pm c$. Show that if f is an even function then $ac = b$. Deduce that if f is an even function then $f(x)$ must reduce to the form $f(x) = k$, where k is constant.

Find all odd functions of the form $\dfrac{ax + b}{x + c}$. (JMB)

13 Sketch the curve given by the equation $y = \dfrac{1 - 2x}{1 - x}$ and shade the region for which

$0 \leqslant x \leqslant \frac{1}{2}, 0 \leqslant y \leqslant 1$ and $y \leqslant \dfrac{1 - 2x}{1 - x}$. (AEB)

14 The functions f and g are defined over the real numbers by

$$f: x \mapsto 2 - x,$$

$$g: x \mapsto \frac{2}{x} \quad (x \neq 0).$$

Show that each of the functions is its own inverse and verify that the inverse of f o g is g o f.

Show that there are no invariant values of f o g. (AEB)

15 Find the coordinates of any turning points and the equations of any asymptotes of the curve

$$y = \frac{x^2 + 1}{x^2 - 4}.$$

Show that the curve has no point of inflexion. Sketch the curve. (AEB)

16 Find the coordinates of the point of intersection of the curve

$$y = \frac{x^3 + 1}{x}$$

and the x-axis. Find also the gradients when $x = 1$ and $x = -1$, the value of x at the stationary point, and the nature of this stationary point. Sketch this curve and also the curve

$$y = \frac{x}{x^3 + 1}. \qquad \text{(L)}$$

17 Find the set of values of y for which x is real when

$$y = \frac{15 + 10x}{4 + x^2}.$$

Find also the gradient of this curve when $x = 0$.
Sketch the curve. (L)

18 Show that for real values of x the expression $\dfrac{2x}{x^2 + 4x + 3}$ cannot lie between certain values. Find these values. Sketch the curve

$$y = \frac{2x}{x^2 + 4x + 3},$$

determining the stationary values of y and the asymptotes of the curve. Show, or confirm, that the curve has a point of inflexion for $x = 3^{2/3} + 3^{1/3}$. (JMB)

19 Show that the range of values taken by the function

$$\frac{x^2 + x + 7}{x^2 - 4x + 5}$$

for real values of x is from $\frac{1}{2}$ to $13\frac{1}{2}$ inclusive. Sketch the graph of the function, indicating its asymptote. (JMB)

20 State the maximum domain and determine the range of the function $f(x)$, where $f(x) \equiv \dfrac{x^2 + 2x + 3}{x^2 - 2x - 3}$. Sketch the curve $y = \dfrac{x^2 + 2x + 3}{x^2 - 2x - 3}$ indicating clearly any asymptotes. (O)

21 Find the equation of the tangent to the parabola $y^2 = \frac{1}{2}x$ at the point $(2t^2, t)$.

Hence, or otherwise, find the equations of the common tangents of the parabola $y^2 = \frac{1}{2}x$ and the ellipse $5x^2 + 20y^2 = 4$.

Draw a sketch to illustrate the results. (O)

22 Sketch the curve $y = \dfrac{x^2 - 7x + 10}{x - 6}$, indicating clearly any vertical asymptotes and

turning values. (O)

23 Draw a sketch-graph of the curve whose equation is

$$y = x^2(2 - x).$$

Hence, or otherwise, draw a sketch-graph of the curve whose equation is

$$y^2 = x^2(2 - x),$$

indicating *briefly* how the form of the curve has been derived. (C)

24 Sketch the graph of the function $f: x \mapsto |x + 1| + |x - 2|$, where $x \in \mathbb{R}$, the set of real numbers.

State the range of the function. (C)

25 The mapping g is defined by

$$g: x \mapsto \frac{2x + 1}{x - 1} \quad (x \in \mathbb{R}, x \neq 1).$$

State the range of g and sketch the graph of $y = g(x)$.
Define the mapping g^{-1}. (C)

26 A curve is given by the parametric equations

$$x = t^2 + 3, \quad y = t(t^2 + 3).$$

(i) Show that the curve is symmetrical about the x-axis.
(ii) Show that there is no part of the curve for which $x < 3$.
(iii) Find $\dfrac{dy}{dx}$ in terms of t, and show that $\left(\dfrac{dy}{dx}\right)^2 \geq 9$.

Using the results (i), (ii) and (iii), sketch the curve. (C)

27 The domain of a mapping m is the set of all real numbers except 1, and the mapping is given by

$$m: x \mapsto \left|\frac{x}{x - 1}\right|.$$

(i) State with a reason whether or not m is a function.
(ii) State the range of m.

24 *Curve sketching*

(iii) Show that
$$[x > 1] \Rightarrow [m(x) > 1] \text{ and that } [x < 0] \Rightarrow [m(x) < 1].$$
(iv) Sketch the graph of $y = m(x)$. \hfill (C)

28 Find the equations of the asymptotes of the curve
$$y = (2x^2 - 2x + 3)/(x^2 - 4x).$$

Using the fact that x is real, show that y cannot take any value between -1 and $5/4$.

Sketch the curve, showing its stationary points, its asymptotes and the way in which the curve approaches its asymptotes. \hfill (L)

29 Given that the function $y = f(x)$ has a stationary value at $x = a$, show that the value will be a minimum if $d^2y/dx^2 > 0$ at $x = a$ and a maximum if $d^2y/dx^2 < 0$ at $x = a$.

Show that the graph of the function
$$y = x^5 - 16(x - 1)^5$$

has two stationary points only, one at $x = \frac{2}{3}$ and the other at $x = 2$. At each of these points, determine whether y has a maximum or minimum value. Sketch the graph of the function. \hfill (JMB)

30 By suitable changes of axes, show that the equations
$$y = \frac{1}{x^2}, \quad y = \frac{1}{(1 - x)^2}, \quad y = 1 + \frac{1}{x^2}$$

represent the same curve in different positions.

Show graphically, or otherwise, that the equation
$$\frac{1}{(1 - x)^2} = 1 + \frac{1}{x^2}$$

has two real roots, one between 0 and 1, the other between 1 and 2. \hfill (OC)

31 Functions f, g and h, with domains and codomains
$$\mathbb{R}^+ = \{x : x \in \mathbb{R}, x > 0\}$$

are defined as follows:
$$f: x \mapsto 3x^2, \quad g: x \mapsto (1 + x)^{-1/2}, \quad h: x \mapsto (1 + x)/x.$$

Prove that the composite function L defined on $\mathbb{R}^+$ by $L = hgf$ is given by
$$L: x \mapsto 1 + \sqrt{(1 + 3x^2)}.$$

Show that the derived function L' is an increasing function of x on $\mathbb{R}^+$ and sketch its graph, indicating in particular its behaviour for large values of x. \hfill (L)

32 For the curve with equation
$$y = \frac{x}{(x - 2)},$$

find
 (i) the equation of each of the asymptotes,
(ii) the equation of the tangent at the origin.

Sketch the curve, paying particular attention to its behaviour at the origin and as it approaches its asymptotes.

On a separate diagram sketch the curve with equation
$$y = \left| \frac{x}{x - 2} \right|. \hfill (L)$$

33 Determine the set of values of λ for which the equation

$$\lambda(x^2 - x + 1) = x^2 + x + 1$$

has real roots.

Sketch the graph of

$$y = \frac{x^2 + x + 1}{x^2 - x + 1}$$

indicating clearly the asymptote and how the curve approaches its asymptote.

Sketch, on a separate diagram, the curve whose equation is

$$y = \frac{x^2 - x + 1}{x^2 + x + 1}. \tag{L}$$

34 Sketch the curve with parametric equations

$$3x = t^3 - 3t, \quad 3y = t^3. \tag{L}$$

35 Sketch the curve $3ay^2 = x^2(2x + a)$, where $a > 0$, and show that the two tangents to this curve at the origin intersect at an angle of $\pi/3$. (L)

36 (a) A function f is defined on the interval $[0, 2]$ as follows:

$$\begin{aligned}
f(x) &= 0 & &\text{if } x = 0, \\
&= x + 1 & &\text{if } x \in (0, 1), \\
&= 2 - x & &\text{if } x \in (1, 2), \\
&= 1 & &\text{if } x = 2.
\end{aligned}$$

Sketch the graph of f, showing clearly the point of the graph at any discontinuity.

Sketch the graph of the function f ∘ f.

State what property f has which ensures that an inverse function exists, and sketch the graph of f^{-1}.

(b) The functions g_1, g_2, g_3 are given by

$$\begin{aligned}
g_1(x) &= |x + 2| - |x - 2|, \\
g_2(x) &= x^4 \sin x + x^2 \cos x, \\
g_3(x) &= (x^2 + 1)\cosh x.
\end{aligned}$$

Classify these functions as even, odd or neither, giving reasons for your answers.(W)

37 Given that $y = x^2 + \dfrac{8}{1 - x}$ ($x \in \mathbb{R}$, $x \neq 1$), find $\dfrac{dy}{dx}$ and $\dfrac{d^2y}{dx^2}$.

Verify that there is a turning point on the graph of y when $x = -1$, and find whether it is a maximum point or a minimum point. Show that there are no other turning points.

Find the coordinates of the point of inflexion on the graph of y.

Sketch the graph of y, showing the turning point and the point of inflexion, and indicate how the graph lies in relation to the curve $y = x^2$ for large positive and negative values of x. (C)

38 Find the points of the curve

$$3x^2 + 2xy + 3y^2 - 4x + 4y - 20 = 0$$

at which the tangents are
(i) parallel to the x-axis,
(ii) parallel to the y-axis.

Prove that the curve is completely contained within the rectangle formed by these tangents. (O)

39 For the curve whose equation is

$$y^2 - 2xy + x^3 = 0,$$

show that the mid-points of all chords parallel to the axis of y lie on the line $y = x$.
Sketch the curve, making clear the form of the curve near the origin. (C)

40 Given that

$$x^4 + x^2y + y^2 = 3,$$

find, in terms of a and b, equations to give the values of $\dfrac{dy}{dx}$ and $\dfrac{d^2y}{dx^2}$ when $x = a$, $y = b$, and hence find the turning values of y and distinguish between them.
Sketch the curve given by the equation. (C)

2 Transcendental curves

All functions of x which are not completely algebraic are called transcendental functions. The graphs and properties of these functions are developed in this chapter.

2.1 The circular functions

In elementary trigonometrical work, angles are measured in degrees. Later a new measure is used; this is called *circular measure* and angles are measured in *radians*. The circular measure of an angle is defined as the ratio of the length of the arc by which it is subtended at the centre of a circle to the length of the radius of that circle. The angle subtended at the centre of a circle by an arc equal in length to the radius of the circle is defined to be 1 *radian*. In particular, $360° \equiv 2\pi$ radians.

Unless stated otherwise, you should assume that angles are measured in radians in all the subsequent work of this chapter. Should you need to evaluate any specific values of a circular function, most modern electronic calculators provide a choice of degrees or radians.

Parts of the graphs of the six basic circular functions, $\sin x$, $\cos x$, $\text{cosec } x$, $\sec x$, $\tan x$ and $\cot x$ are shown in Fig. 2.1(a), (b), (c), (d), (e) and (f) respectively.

The functions $\sin x$, $\text{cosec } x = \dfrac{1}{\sin x}$, $\tan x = \dfrac{\sin x}{\cos x}$ and $\cot x = \dfrac{\cos x}{\sin x}$ are all odd functions and have point symmetry about O; i.e., $f(x) = -f(-x)$. The functions $\cos x$ and $\sec x = \dfrac{1}{\cos x}$ are even functions and are symmetrical about the y-axis; i.e. $f(x) = f(-x)$.

The functions $\sin x$, $\cos x$, $\text{cosec } x$ and $\sec x$ are periodic and have period 2π; i.e. $f(x) = f(x + 2\pi k)$ for all $k \in \mathbb{Z}$.

The functions $\tan x$ and $\cot x$ are periodic and have period π; i.e. $f(x) = f(x + k\pi)$ for all $k \in \mathbb{Z}$.

Sine waves

The function $f: x \mapsto a \sin x + b \cos x + c$ may be rewritten as

$$f: x \mapsto R \cos (x - \alpha) + c,$$

where $R = \sqrt{(a^2 + b^2)}$, taking the positive square root only, and $\cos \alpha = b/\sqrt{(a^2 + b^2)}$ and $\sin \alpha = a/\sqrt{(a^2 + b^2)}$, giving a unique value of α in the interval $0 \leqslant \alpha < 2\pi$.

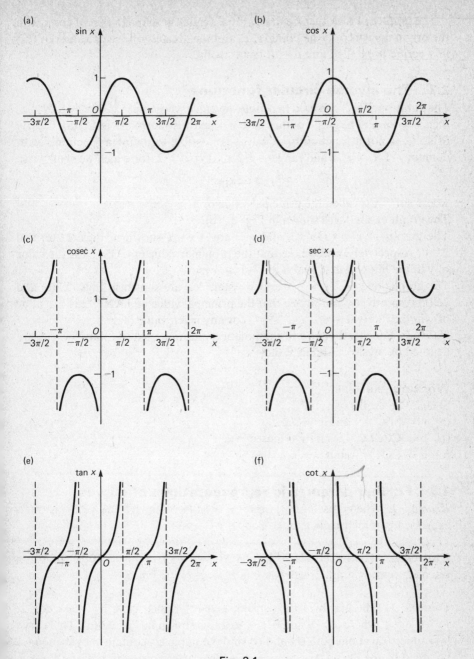

Fig. 2.1

The graph of f may then be drawn; it is similar in shape to that of cos x with the origin displaced to the point (α, c) and the greatest and least values of $f(x)$ in a period are $c + R$ and $c - R$ respectively.

2.2   The inverse circular functions

The notation $\text{SIN}^{-1} x$ is used to denote 'an angle whose sin is x'. If $y = \text{SIN}^{-1} x$, then $x = \sin y$. The graph of the curve $x = \sin y$ is shown in Fig. 2.2(a). Clearly $\text{SIN}^{-1} x$ is not a function of x, but it is possible to define a function f with domain $-1 \leqslant x \leqslant 1$ and range $-\pi/2 \leqslant f(x) \leqslant \pi/2$, for which we shall write

$$\text{f}: x \mapsto \sin^{-1} x,$$

where $\sin^{-1} x$ is called the principal value of $\text{SIN}^{-1} x$.
The graph of $\sin^{-1} x$ is shown in Fig. 2.2(b).
The graphs of $y = \text{COS}^{-1} x$ and $y = \cos^{-1} x$ are shown in Figs. 2.2(c) and 2.2(d) respectively, and we see that the principal value of $\text{COS}^{-1} x$ is the value of y in the interval $0 \leqslant y \leqslant \pi$ for $-1 \leqslant x \leqslant 1$.
The graphs of $y = \text{TAN}^{-1} x$ and $y = \tan^{-1} x$ are shown in Figs. 2.2(e) and 2.2(f) respectively, and we see that the principal value of $\text{TAN}^{-1} x$ is the value of y in the interval $-\pi/2 \leqslant y \leqslant \pi/2$ for any given value of x.
Note: In some textbooks and examination questions, $\sin^{-1} x$, $\tan^{-1} x$, etc. are denoted by arcsin x, arctan x, etc.

Working exercise
Sketch the graphs of
(a) $y = \text{SEC}^{-1} x$ and $y = \sec^{-1} x$,
(b) $y = \text{COSEC}^{-1} x$ and $y = \text{cosec}^{-1} x$,
(c) $y = \text{COT}^{-1} x$ and $y = \cot^{-1} x$.

2.3   Further parametric representations of curves
Example 1   The ellipse $b^2 x^2 + a^2 y^2 = a^2 b^2$ (see Fig. 1.10(a), p. 19) is often expressed parametrically as $x = a \cos t$, $y = b \sin t$.

Example 2   The hyperbola $b^2 x^2 - a^2 y^2 = a^2 b^2$ (see Fig. 1.10(b), p. 19) is often expressed parametrically as $x = a \sec t$, $y = b \tan t$.

Example 3   The graph of the astroid, given parametrically by $x = a \cos^3 t$, $y = a \sin^3 t$, where $a > 0$ and $0 \leqslant t < 2\pi$, is shown in Fig. 2.3(a). This curve is symmetrical about both Ox and Oy and is completely contained by the square bounded by the lines $x = \pm a$, $y = \pm a$. The cartesian equation of the curve is $x^{2/3} + y^{2/3} = a^{2/3}$.
For this curve,

$$\frac{dy}{dx} = \frac{3a \sin^2 t \cos t}{-3a \sin t \cos^2 t} = -\tan t$$

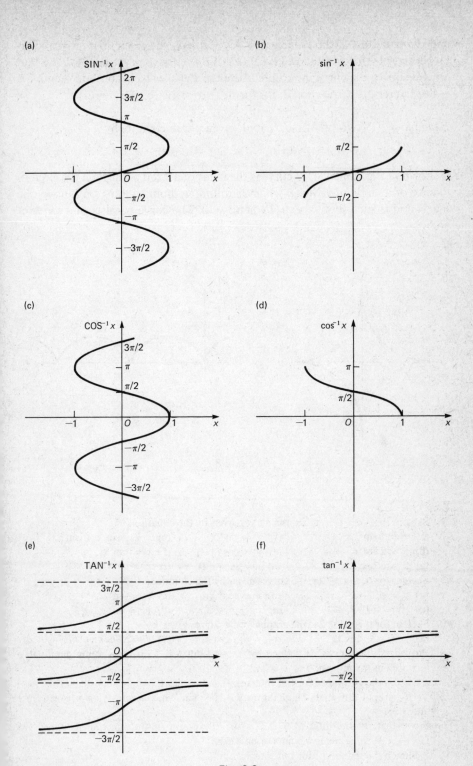

Fig. 2.2

and the equation of the tangent at t is given by $x \sec t + y \operatorname{cosec} t = a$. This tangent meets the coordinate axes at $(a \cos t, 0)$, and $(0, a \sin t)$. It follows that all tangents to the curve intercepted between the axes have constant length a.

See Exercise 2.3, question 8, for further properties of the curve.

Example 4 The graph of the cycloid, given parametrically by

$$x = a(t - \sin t), \quad y = a(1 - \cos t), \quad a > 0,$$

is shown in Fig. 2.3(b). This curve is the locus of a fixed point P on the circumference of a circle of radius a which is rolling without slipping on (above) the x-axis and is such that P is at O when $t = 0$. The curve consists of a series of arches.

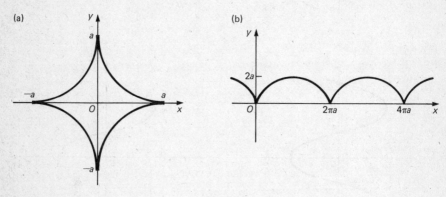

Fig. 2.3

Exercise 2.3

1 Sketch, for $-2\pi \leqslant x \leqslant 2\pi$, the curves given by the equations
   (a) $y = 2 \sin x$,   (b) $y = \sin 2x$,   (c) $y = 2 \sin x + \sin 2x$,
   (d) $y = 2 \sin x - \sin 2x$,   (e) $y = \cos (x/2)$,   (f) $y = \tan 3x$,
   (g) $y = \operatorname{cosec} 2x$,   (h) $y = \sec (x/2)$,   (i) $y = \cot (x/3)$.

2 Sketch, for $0 \leqslant x \leqslant 2\pi$, the curves given by the equations
   (a) $y = \sin^2 x$,   (b) $y = \cos^2 x$,   (c) $y = \tan^2 x$,
   (d) $y^2 = \sin x$,   (e) $y^2 = \operatorname{cosec} x$,   (f) $y^2 = \cot x$.

3 Sketch, for $0 \leqslant x \leqslant 2\pi$, the graphs of the curves given by
   (a) $y = x + \cos x$,   (b) $y = x \cos x$,   (c) $y = x/(\cos x)$.

4 Prove that the function f defined by $f : x \mapsto \frac{1}{2} \tan x - 4 \sin x$ has a minimum value $-3\sqrt{3}/2$ in the interval $0 \leqslant x \leqslant \pi/2$.
   Sketch the graph of f for this interval.

5 Verify graphically that the equation $\theta - 1 = \sin \theta$ has a root lying between 1·90 and 1·95.

6 Prove that, in the interval $0 \leqslant x \leqslant 4\pi$, the graph of the function f defined by $f : x \mapsto x - \sin x$ has five points of inflexion.
   Sketch the graph of f for this interval.

7 Sketch, for $-\pi \leqslant x \leqslant \pi$, the curves given by the equations
   (a) $y = 3 \sin x + 4 \cos x$, (b) $y = 3 \sin x - 4 \cos x$, (c) $y = 9 \sin^2 x - 16 \cos^2 x$.
8 For the astroid, shown in Fig. 2.3(a) (see also Example 3, p. 30), prove that
   (a) the area enclosed by the curve is $3\pi a^2/8$,
   (b) the length of the perimeter of the curve is $6a$.
9 Sketch the curves, for $a > 0$, given parametrically by the equations
   (a) $x = a(2 \cos t - \cos 2t)$, $y = a(2 \sin t - \sin 2t)$,
   (b) $x = a(2 \cos t + \cos 2t)$, $y = a(2 \sin t - \sin 2t)$,
   (c) $x = a(3 \cos t - \cos 3t)$, $y = a(3 \sin t - \sin 3t)$.
   Hints: (a) is a closed curve, symmetrical about the x-axis, with a cusp at $(a, 0)$; it is called a *cardioid*;
   (b) is a closed curve with cusps at points corresponding to $t = 0$, $2\pi/3$ and $4\pi/3$; it is symmetrical about the lines $x = 0$, $y = \pm x\sqrt{3}$; it is called a *deltoid*;
   (c) is a closed curve, symmetrical about both the x- and y-axes, with cusps at $(\pm 2a, 0)$; it is called a *nephroid*.
10 Use the inequality, $\sin \theta < \theta < \tan \theta$ for $0 < \theta < \pi/2$, to show, in a single sketch, the relation between the graphs of $y = \sin x$, $y = x$ and $y = \tan x$ in the interval $-\pi/2 < x < \pi/2$.
11 Write down the values of
   (a) $\tan^{-1} 1$, (b) $\sin^{-1}(-\tfrac{1}{2})$, (c) $\cos^{-1}(-1)$, (d) $\operatorname{cosec}^{-1}(-2)$,
   (e) $\sec^{-1}(2\sqrt{3}/3)$, (f) $\cot^{-1}(-\sqrt{3})$.
12 Sketch, for $-1 \leqslant x \leqslant 1$, the graphs of the curves given by
   (a) $y = 5 \sin^{-1} x$, (b) $y = 3 \cos^{-1} x$, (c) $y = 5|\sin^{-1} x|$.

2.4 Exponential functions

In mathematics, the word exponent means the same as *index* and functions such as

$$\text{f}: x \mapsto 10^x, \qquad \text{g}: x \mapsto 2^{\sin x},$$

where the index varies, are called exponential functions. In books on calculus, it is shown that the exponential function given by $\text{f}(x) = a^x$, where $a > 1$, has a derived function f', given by $\text{f}'(x) = ka^x$, where k is a positive constant, dependent on the value of a.

Writing $y = a^x$, the result is succinctly expressed by the differential equation $\dfrac{\mathrm{d}y}{\mathrm{d}x} = ky$, and such equations provide a satisfactory model for the behaviour of many physical phenomena having rates of growth or decay which depend on y.

Further, as a takes any specific value from the set of real numbers which are greater than 1, a unique value of k from the set of real positive numbers will correspond to this specific value of a, and conversely. By choosing $k = 1$ we are able to identify a unique value of a, called e, and for this function $y = \mathrm{e}^x$ and $\dfrac{\mathrm{d}y}{\mathrm{d}x} = \mathrm{e}^x$.

The number e is irrational and its value is $2{\cdot}718\,28 \ldots$.

The function e^x is called *the exponential function* and its graph is shown in Fig. 2.4(a).

Working exercise

On the same axes, sketch the graphs of $y = 2^x$, $y = e^x$ and $y = 3^x$.

Note that all curves $y = a^x$ pass through the point $(0, 1)$. The curve $y = 2^x$ lies below that of $y = e^x$ for $x > 0$ and the curve $y = 2^x$ lies above $y = e^x$ for $x < 0$; the curve $y = 3^x$ lies above that of $y = e^x$ for $x > 0$ and below for $x < 0$. The graph of the function e^{-x} is obtained by reflecting the graph of $y = e^x$ in the y-axis and is shown in Fig. 2.4(b).

The graph of the functions $y = e^{x^2}$ and $y = e^{-x^2}$ are shown in Figs. 2.4(c) and (d), respectively.

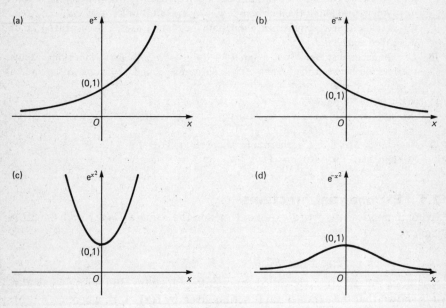

Fig. 2.4

2.5 Hyperbolic functions

There are certain combinations of exponential functions which have properties with a close analogy to those of the circular functions. They are called the hyperbolic sine, the hyperbolic cosine, etc. and are *defined* and denoted as:

$$\sinh x = \tfrac{1}{2}(e^x - e^{-x}), \quad \operatorname{cosech} x = \frac{1}{\sinh x} = \frac{2}{e^x - e^{-x}},$$

$$\cosh x = \tfrac{1}{2}(e^x + e^{-x}), \quad \operatorname{sech} x = \frac{1}{\cosh x} = \frac{2}{e^x + e^{-x}},$$

$$\tanh x = \frac{\sinh x}{\cosh x} = \frac{e^x - e^{-x}}{e^x + e^{-x}} = \frac{1 - e^{-2x}}{1 + e^{-2x}} = \frac{e^{2x} - 1}{e^{2x} + 1},$$

$$\coth x = \frac{1}{\tanh x} = \frac{e^x + e^{-x}}{e^x - e^{-x}} = \frac{1 + e^{-2x}}{1 - e^{-2x}} = \frac{e^{2x} + 1}{e^{2x} - 1}.$$

We note that $\cosh x$ and $\operatorname{sech} x$ are even functions and $\sinh x$, $\operatorname{cosech} x$, $\tanh x$ and $\coth x$ are odd functions.

The graphs of $y = \sinh x$, $y = \cosh x$ and $y = \tanh x$ are shown in Fig. 2.5(a) and the graphs of their inverse relations $y = \sinh^{-1} x$, $y = \text{COSH}^{-1} x$ and $y = \tanh^{-1} x$ are shown in Fig. 2.5(b).

In some books the notation arsinh x, arcosh x, artanh x, etc. is used for $\sinh^{-1} x$, $\cosh^{-1} x$, $\tanh^{-1} x$, etc. respectively.

Note that the graphs in Fig. 2.5(b) are obtained by reflecting those in Fig. 2.5(a) in the line $y = x$. The graphs show that $\text{COSH}^{-1} x$ is two-valued, but that $\sinh^{-1} x$ and $\tanh^{-1} x$ are functions of x.

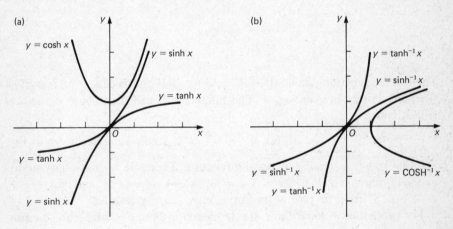

Fig. 2.5

Working exercise

Sketch the graphs of $y = \operatorname{cosech} x$, $y = \operatorname{sech} x$ and $y = \coth x$.

Use your graphs to sketch the graphs of the inverse relations $y = \operatorname{cosech}^{-1} x$, $y = \text{SECH}^{-1} x$ and $y = \coth^{-1} x$.

Note: The curve $y = c \cosh(x/c)$ is called a *catenary* and is the form assumed by a uniform flexible chain suspended between two points and hanging in a vertical plane under gravity.

2.6 Logarithm functions

A scrutiny of the graph of the exponential function (see Fig. 2.4(a)) suggests that, for all $x \in \mathbb{R}$, the function

$$f: x \mapsto e^x$$

is one–one. The graph of the inverse function f^{-1} is obtained by reflecting the graph of f in the line $y = x$, as shown in Fig. 2.6.

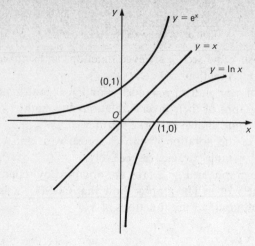

Fig. 2.6

The inverse function has domain $\mathbb{R}^+$ and range $\mathbb{R}$ and is called *the logarithm function*, written $\ln x$ or $\log_e x$. The number e is called the base of natural logarithms.

If $y = \ln x$, $\dfrac{dy}{dx} = \dfrac{1}{x}$ and, for $x > 0$, $\dfrac{dy}{dx} > 0$, confirming, as shown in Fig. 2.6, that $\ln x$ is a steadily·increasing function. There are no stationary points on the graph of $\ln x$. When $x = 1$, $y = 0$; as $x \to \infty$, $y \to \infty$, and as $x \to +0$, $y \to -\infty$. There is no part of the curve for which x is negative.

The properties of logarithms are developed in books on algebra. The most important are listed here:

If $a^x = N$, then
$$x = \log_a N,$$
$$\log_a M + \log_a N = \log_a (MN),$$
$$\log_a M - \log_a N = \log_a (M/N),$$
$$\log_a (M^k) = k \log_a M,$$
$$\log_a M = \log_b M/(\log_b a).$$

The following limits are also important:

(i) When $x \to +\infty$, for all α, $e^x/x^\alpha \to +\infty$ and $e^{-x}x^\alpha \to 0$,

(ii) When $x \to +\infty$, for $\alpha > 0$, $(\ln x)/x^\alpha \to 0$,

(iii) When $x \to +0$, for $\alpha > 0$, $x^\alpha \ln x \to 0$.

Example 5 Sketch the curve given by $y = \log_{10} x$.

Since $y = \log_{10} x = \ln x/(\log_e 10) = \log_{10} e \cdot \ln x$, the graph of $y = \log_{10} x$ is similar to that of $\ln x$, with scale factor $\log_{10} e$.

Example 6 Sketch the curve given by $y = \ln x^k$, where
(a) k is an odd positive integer, (b) k is an even positive integer.

(a) In this case, the graph of $y = \ln x^k$ is similar to that of $y = \ln x$, with scale factor k. If $x < 0$, $x^k < 0$ and no part of the graph exists to the left of the y-axis (see Fig. 2.7(a)).

(b) In this case, we note that

$$\text{for } x > 0, \quad y = k \ln x,$$

$$\text{for } x < 0, \quad y = k \ln (-x).$$

The graph of $y = \ln x^k$ consists of two branches which are symmetrical about the y-axis (see Fig. 2.7(b)).

Example 7   Sketch the curve given by $y = x^2 e^{-x^2}$.

Since $x^2 e^{-x^2}$ is an even function, the curve is symmetrical about the y-axis.

$$\frac{dy}{dx} = 2x(1 - x^2)e^{-x^2} \Rightarrow \frac{dy}{dx} = 0 \quad \text{when } x = 0, \pm 1.$$

$$\frac{d^2y}{dx^2} = 2(1 - 5x^2 + 2x^4)e^{-x^2}.$$

x	y	$\dfrac{dy}{dx}$	$\dfrac{d^2y}{dx^2}$	Nature of stationary point
0	0	0	2	Minimum
1	e^{-1}	0	$-4e^{-1}$	Maximum
-1	e^{-1}	0	$-4e^{-1}$	Maximum

As $x \to \pm\infty$, $y \to 0$.

Since $\dfrac{d^2y}{dx^2} = 0$ when $x = \pm\sqrt{[\frac{1}{4}(5 \pm \sqrt{17})]}$, there are four points of inflexion.

The curve is shown in Fig. 2.7(c).

Example 8   Sketch, for $x \geqslant 0$, the curve given by $y = e^{-x} \sin x$.

Since $-1 \leqslant \sin x \leqslant 1$ and $e^{-x} > 0$,
$-e^{-x} \leqslant e^{-x} \sin x \leqslant e^{-x}$ with equality occurring for the left-hand inequation when $x = (2n - \frac{1}{2})\pi$ and for the right-hand inequation when $x = (2n + \frac{1}{2})\pi$, where $n \in \mathbb{Z}^+$.

This implies that the curves $y = -e^{-x}$ and $y = e^{-x} \sin x$ and the curves $y = e^{-x}$ and $y = e^{-x} \sin x$ have common points. Further, because of the inequalities, $y = e^{-x} \sin x$ cannot lie below $y = -e^{-x}$ or above $y = e^{-x}$.

Also, each of the curves has tangents at every point, and so they must touch. However, since $y = e^{-x}$ and $y = -e^{-x}$ have negative and positive gradient, respectively, at every point, their common tangents with the curve $y = e^{-x} \sin x$ will never occur at the maxima and the minima of $y = e^{-x} \sin x$.

A sketch of the curve $y = e^{-x} \sin x$ is shown in Fig. 2.7(d).

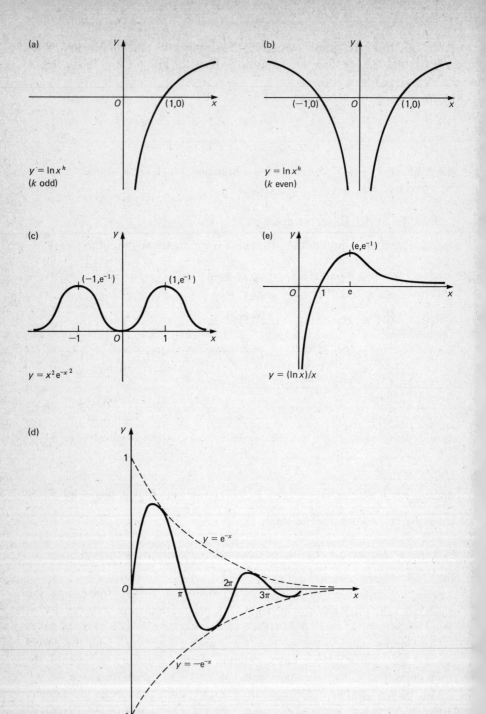

(a)

$y = \ln x^k$
(k odd)

(b)

$y = \ln x^k$
(k even)

(c)

$(-1, e^{-1})$ $(1, e^{-1})$

$y = x^2 e^{-x^2}$

(e)

(e, e^{-1})

$y = (\ln x)/x$

(d)

$y = e^{-x}$

2π

3π

$y = -e^{-x}$

$y = e^{-x} \sin x$ for $x \geqslant 0$

Fig. 2.7

Note: A similar argument may be used to sketch the curves given by $y = x \sin x$, $y = (\sin x)/x$, etc. See also Chapter 3, Fig. 3.3(c), p. 49.

Example 9   Sketch the curve given by $y = (\ln x)/x$.

No part of the curve exists for $x < 0$.
As $x \to +0$, $y \to -\infty$.
As $x \to \infty$, $y \to +0$.
When $x = 1$, $y = 0$.

$$\frac{dy}{dx} = (1 - \ln x)/x^2 \quad \text{and} \quad \frac{dy}{dx} = 0 \quad \text{when} \quad x = e.$$

$$\frac{d^2y}{dx^2} = (2 \ln x - 3)/x^3 \quad \text{and} \quad \frac{d^2y}{dx^2} = 0 \quad \text{when} \quad x = e^{3/2}.$$

The point (e, e^{-1}) is a maximum, the point $(e^{3/2}, \frac{3}{2}e^{-3/2})$ is a point of inflexion. The curve is shown in Fig. 2.7(e).

Exercise 2.6

1   Sketch the curves given by the equations
(a) $y = e^{2x}$,   (b) $y = e^{x/2}$,   (c) $y = e^{1/x}$,
(d) $y = e^{1/x^2}$,   (e) $y = e^{-2x}$,   (f) $y = e^{-1/x^2}$.

2   Use a graphical method to show that the equation $e^x = x + a$ has no real roots if $a < 1$ and two roots if $a > 1$.

3   Using the same axes, sketch, for $0 < x < 2\pi$, the curves given by the equations $y = \sin x$, $y = \cos x$, $y = e^{\sin x}$ and $y = e^{\cos x}$.

4   (a) Prove that the curve $y = xe^x$ takes its minimum value at the point $(-1, -e^{-1})$ and sketch the curve.
(b) Prove that the curve $y = x^2e^x$ takes stationary values at the points $(0, 0)$ and $(-2, 4e^{-2})$ and sketch the curve.

5   Using the same axes, sketch, for $-\pi/2 < x < 3\pi/2$, the curves $y = \tan x$, $y = e^{\tan x}$ and $y = e^{-\tan x}$.

6   Sketch the curves given by the equations
(a) $y = \ln x$,   (b) $y = \ln (2x)$,   (c) $y = \ln (x^2)$,
(d) $y = \ln (x - 2)$, (e) $y = \ln |x|$,   (f) $y = \ln |x - 2|$.

7   Sketch the graph of the function f given by

$$f: x \mapsto (1 + 2x^2)e^{-x^2}.$$

(*Hint:* Show that the function has three stationary values.)

8   Show that the curve $y = \ln [\tan (x/2 + \pi/4)]$ has a point of inflexion at the origin. Prove that $\sinh y = \tan x$ and $\cosh y = \sec x$ in the interval $-\pi/2 < x < \pi/2$. Sketch the curve in this interval.

9   For the function f: $x \mapsto a \cosh x + b \sinh x$, where a and b are both positive constants, prove that
(a) if $a > b$, f has a minimum value $\sqrt{(a^2 - b^2)}$,
(b) if $a < b$, f has no stationary values.
Sketch the graph of f for (i) $a > b$,   (ii) $a < b$.

10   Sketch, in the same diagram, the graphs of $y = e^x$, $y = e^{-x}$, $y = \cosh x$, $y = \sinh x$ and $y = \operatorname{arcosh} x$, showing their correct relative positions.

11 Sketch, for $-\pi/2 < x < \pi/2$, the curve $y = \ln (\sec x)$.

12 Sketch, for $-2\pi < x < 2\pi$, the curves $y = e^x \sin^2 x$ and $y = e^{-x} \sin^2 x$.

13 Sketch the curves whose equations are
    (a) $y = \ln (1 - x^2)$,   (b) $y = \ln (x^2 - 1)$,   (c) $y = \ln (1 + x^2)$.

14 Sketch the graph of $y = \ln (x^2 + x^{-2})$ and show that the curve has two points of inflexion.

15 Sketch the curves given by the equations
    (a) $y = x \ln x$,   (b) $y = x^2 \ln x$,
    (c) $y = x \ln x^2$,   (d) $y = x^2 \ln x^2$.

Miscellaneous Exercise 2

1 Sketch the curve $y = 1 + \sin 2x$ for $0 \leqslant x \leqslant \pi$.   (L)

2 Sketch, in a single diagram, graphs of the functions f and g given that

$$f: x \mapsto 2 \cos x, \quad \text{where } x \in \mathbb{R} \text{ and } -\pi/2 \leqslant x \leqslant \pi/2,$$

$$g: x \mapsto e^x, \quad \text{where } x \in \mathbb{R} \text{ and } -\pi/2 \leqslant x \leqslant \pi/2.$$

Write on your diagram the coordinates of each of the points where the curves meet the coordinate axes.   (L)

3 Sketch a graph of the function

$$f: x \mapsto 2^x \quad \text{where } x \in \mathbb{R}.$$

By adding a straight line to your sketch, find the number of real roots of the equation

$$x + 2^x = 2.$$

State the equation of the line you have drawn.   (L)

4 Given that $x \in \mathbb{R}$, sketch the graph of the function f defined by

$$f: x \mapsto \begin{cases} \log_2 x & \text{if } x > 0, \\ -\log_2(-x) & \text{if } x < 0. \end{cases} \quad\quad \text{(L)}$$

5 If $y = \sin x \cos^3 x \ (0 \leqslant x \leqslant \pi)$, find the values of x for which $\dfrac{dy}{dx} = 0$, and sketch the graph of y against x for $0 \leqslant x \leqslant \pi$.   (C)

6 The parametric equations of a curve are

$$x = 4 \sinh u, \quad y = \cosh 2u.$$

Sketch the curve.   (JMB)

7 The function f is periodic with period π and

$$f(x) = \sin x \quad \text{for } 0 \leqslant x \leqslant \pi/2,$$

$$f(x) = 4(\pi^2 - x^2)/(3\pi^2) \quad \text{for } \pi/2 < x < \pi.$$

Sketch the graph of $f(x)$ in the range $-\pi \leqslant x \leqslant 2\pi$.   (L)

8 Show that the only stationary points on the graph of the function $y = 2 \cos 3x - \cos 2x + 2 \cos x$, in the range $-\pi/2 \leqslant x \leqslant \pi/2$, occur at $x = -\pi/3$, $x = 0$ and $x = \pi/3$ and determine whether each is a maximum, a minimum or a point of inflexion.
    Find the values of x at which $y = 0$.
    Sketch the graph of y in this range of x.   (JMB)

9 Given that $y = \sin 2x + 2 \cos x$, find the values of x in $[0, 2\pi]$ for which $\dfrac{dy}{dx} = 0$.

For each such value, determine whether the corresponding value of y is a maximum, a minimum or neither. Sketch the graph of y against x for $0 \leqslant x \leqslant 2\pi$. $\qquad$ (C)

10 Sketch the curve whose equation is

$$y = \ln (1 - 3x),$$

for $x < \frac{1}{3}$. $\qquad$ (L)

11 Functions g and h are defined on the set of real numbers by

$$g(x) = 2 - \cos x,$$

$$h(x) = 2 + \sin \frac{2x}{3}.$$

For each function, state (a) the period, (b) whether the function is odd, even or neither. Sketch graphs of these two functions for

$$-\frac{3\pi}{2} < x < \frac{3\pi}{2}. \qquad \text{(L)}$$

12 A curve joining the points $(0, 1)$ and $(0, -1)$ is represented parametrically by the equations

$$x = \sin \theta, \quad y = (1 + \sin \theta) \cos \theta,$$

where $0 \leqslant \theta \leqslant \pi$. Find dy/dx in terms of θ, and determine the x,y coordinates of the points on the curve at which the tangents are parallel to the x-axis and of the point at which the tangent is perpendicular to the x-axis. Sketch the curve. $\qquad$ (O)

13 Find the minimum value of $x \ln x$, where x is positive. Given that $x \ln x$ tends to 0 as x tends to 0, sketch the curve $y = x \ln x$. $\qquad$ (L)

14 Show that the curve

$$y = (1 + x)e^{-2x}$$

has just one maximum and one point of inflexion.

Sketch the curve, marking the coordinates of the maximum point, the point of inflexion and the point where the curve crosses the x-axis.

[You may assume that $(1 + x)e^{-2x} \to 0$ as $x \to +\infty$.] $\qquad$ (L)

15 Investigate the stationary values of the function

$$\text{f}: x \mapsto x^2 e^{-ax}, \quad \text{where } a \neq 0 \text{ and } x \text{ is real.}$$

Sketch the graph of the function in the cases (i) $a = -1$, (ii) $a = 2$. $\qquad$ (L)

16 Express $\text{f}(x) \equiv 24 \cos x - 7 \sin x$ in the form $R \cos (x + \alpha)$, where R is positive and α is an acute angle, giving the value of α in degrees and minutes.

State the maximum and minimum values of $\text{f}(x)$.

Find the general solution of the equation $\text{f}(x) = 0$.

Sketch the graph of $y = \text{f}(x)$ for $-360^\circ \leqslant x \leqslant 360^\circ$. $\qquad$ (L)

17 Expand the function

$$\text{f}(x) = \tfrac{1}{2}(x - 1)e^{2x} + \tfrac{1}{2}(x + 1)e^{-2x}$$

in powers of x as far as the term in x^3.

Sketch the curve $y = \text{f}(x)$ in the neighbourhood of the origin, showing the tangent at the origin, and find the equation of this tangent. Indicate on your figure how the curve bends away from the tangent for small positive and small negative values of x. $\qquad$ (OC)

18 Given that $-1 < x < 1$, that $0 < \cos^{-1}x < \pi$ and that $(1 - x^2)^{1/2}$ denotes the positive square root of $1 - x^2$, find the derivative of the function

$$f(x) = \cos^{-1}x - x(1 - x^2)^{1/2},$$

expressing your answer as simply as possible.

Prove that, as x increases in the interval $-1 < x < 1$, $f(x)$ decreases, and sketch the graph of $f(x)$ in this interval. (JMB)

19 By considering the derivative of the function

$$\ln (1 + x) - \frac{x^2}{(1 + x)^2}$$

prove that, if $x > 0$, then $\ln (1 + x) > \dfrac{x^2}{(1 + x)^2}$.

Sketch, on a single diagram, the graphs of the functions. (JMB)

20 If k is a positive constant, show that there is one turning point on the graph of $y = \ln x - kx$ ($x \in \mathbb{R}^+$), and determine whether it is a maximum or a minimum point. Find the value of k for which the turning point is on the x-axis, and sketch the graph in this case. (C)

21 Express the function $4 \cos x - 6 \sin x$ in the form $R \cos (x + \alpha)$ where R is positive and $0° < \alpha < 360°$.

Hence find the coordinates of the point on the graph of $y = 4 \cos x - 6 \sin x$ for $0° \leqslant x \leqslant 360°$ where y has its minimum value.

Sketch the graph. State the values of x where the curve crosses the x-axis. (AEB)

22 Using the same axes sketch the curves given by $y = e^x$ and $y = 3 - 4x - x^2$. If t is the positive root of the equation $e^x = 3 - 4x - x^2$, prove that the area of the finite region in the first quadrant enclosed by the curves and the y-axis is given by $\frac{1}{3}(21t - 3t^2 - t^3 - 6)$. (AEB)

23 Find the values of θ, lying in the interval $-\pi \leqslant \theta \leqslant \pi$, for which $\sin 2\theta = \cos \theta$.

Sketch on the same axes the graphs of $y = \sin 2\theta$ and $y = \cos \theta$ in the above interval and deduce the set of values of θ in this interval for which $\sin 2\theta \geqslant \cos \theta$. (JMB)

24 The functions f, g are defined for $x > 0$ by

$$f: x \mapsto x^2,$$
$$g: x \mapsto \log_e x.$$

Sketch and label the graphs of g, f $\circ$ g and g^{-1} on the same axes, using the same scale. (AEB)

25 The function f is defined for $x \geqslant 0$ by

$$f: x \mapsto 1 + \ln (2 + 3x).$$

Find (i) the range of f,
     (ii) an expression for f$^{-1}(x)$,
     (iii) the domain of f^{-1}.

On one diagram, sketch graphs of f and f^{-1}.

Show that the x-coordinate of the point of intersection of the graphs of f and f^{-1} satisfies the equation

$$\ln (2 + 3x) = x - 1.$$ (JMB)

26 Three functions f, g and h are defined on the domain $-1 \leqslant x \leqslant 1$ as follows:

$$f: x \mapsto \sin^{-1} x \quad \text{and} \quad -\tfrac{1}{2}\pi \leqslant f(x) \leqslant \tfrac{1}{2}\pi;$$

$$g: x \mapsto \cos^{-1} x \quad \text{and} \quad 0 \leqslant g(x) \leqslant \pi;$$
$$h: x \mapsto \tan^{-1} x \quad \text{and} \quad -\tfrac{1}{4}\pi \leqslant h(x) \leqslant \tfrac{1}{4}\pi.$$

Sketch separately the graphs of f, g and h.

If $g(x) = h(x)$, prove that $x^4 + x^2 - 1 = 0$.   (C)

27   Let $f: x \mapsto x e^{-x}$, where x is real. Find the coordinates of the turning point on the graph of $y = f(x)$, and sketch the graph. It may be assumed that $x e^{-x} \to 0$ as $x \to \infty$.

Let $g: x \mapsto x^n e^{-x}$, where n is a positive integer greater than 1 and where x is real. Find the values of x for which $g(x)$ has stationary values, and determine the nature of these stationary values, considering separately the cases n even and n odd.   (C)

28   Sketch the curve with parametric equations $x = 4 \cos t$, $y = \cos 2t$.

The portion of this curve between $t = 0$ and $t = \tfrac{1}{2}\pi$ is rotated completely about the y-axis. Prove that the area of the curved surface thus generated is $\tfrac{32}{3}\pi(2\sqrt{2} - 1)$.   (C)

29   Sketch, for all real values of x, the curve whose equation is

$$y = \operatorname{arsinh} x.$$

Find the angle between the x-axis and the tangent to this curve at the origin, and draw this tangent on your sketch.   (L)

30   Sketch the curve given by

$$x = a \cos^3 \theta, \quad y = b \sin^3 \theta, \quad (0 \leqslant \theta \leqslant 2\pi),$$

where the constants a and b are both positive.

Find $\dfrac{dy}{dx}$ in terms of θ and prove that there are four points on the curve at which the normal passes through the origin.

The region bounded by that portion of the curve for which $0 \leqslant \theta \leqslant \pi$ and the x-axis is rotated through 2π about the x-axis. Prove that the volume of the solid of revolution formed is

$$\frac{32}{105}\pi a b^2. \qquad \text{(C)}$$

31   The curve $y = e^{-ax} \cos(x + b)$ is denoted by S and the curves $y = e^{-ax}$ and $y = -e^{-ax}$ by E_1 and E_2 respectively $(x > 0, a > 0)$. Show that E_1 and E_2 each touch S at every point at which they meet S.   (JMB)

32   A curve is given by the equations

$$x = 2 \cos t - \cos 2t - 1,$$
$$y = 2 \sin t - \sin 2t, \quad (0 \leqslant t < 2\pi).$$

In any order,
(i)   show that the tangents to the curve at the points $t = \tfrac{1}{3}\pi$, $t = \tfrac{2}{3}\pi$, $t = \pi$, $t = \tfrac{4}{3}\pi$ form a rectangle;
(ii)   sketch the curve;
(iii)   prove that the length of the curve is 16;
(iv)   show that the equation of the curve in polar coordinates may be expressed in the form $r = 2(1 - \cos \theta)$.   (C)

33   Show that, for small values of x, the principal value of $\tan^{-1}(1 + x^2)$ is approximately $\tfrac{1}{4}\pi + \tfrac{1}{2}x^2$. Sketch the graph of $y = \tan^{-1}(1 + x^2)$, considering only the principal value. Indicate clearly any asymptotes.   (C)

3 Curve sketching in polar coordinates

3.1 Polar coordinates

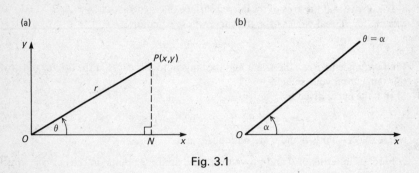

Fig. 3.1

In Fig. 3.1(a), the position of the point P is given by the position vector $\overrightarrow{OP}$ or by the cartesian coordinates (x, y), where $ON = x$, $NP = y$, with the normal sign convention applying when P takes positions in the other quadrants defined by the axes Ox and Oy. The point P may also be given by the coordinates (r, θ), where r is the magnitude of the vector $\overrightarrow{OP}$ and θ is the angle xOP, measured from Ox to OP with the usual sign convention used in trigonometry. The coordinates (r, θ) are called the *polar coordinates* of the point P. The point O is called the *pole*, $\overrightarrow{OP}$ is called the *radius vector* and Ox is called the *initial line*, sometimes called Ol.

In order to ensure that any point $P(x, y)$ in the plane has a unique polar coordinate representation (r, θ), it is necessary and sufficient to define $r \geqslant 0$ and to restrict θ to an interval such as $0 \leqslant \theta < 2\pi$ or $-\pi < \theta \leqslant \pi$. With these restrictions we have:

for given (r, θ), $\qquad x = r\cos\theta, \qquad\qquad y = r\sin\theta;$

and for given (x, y), $\qquad r = +\sqrt{(x^2 + y^2)}, \qquad \theta = \arctan(y/x).$

Unless these restrictions are imposed, r and θ are not necessarily unique in terms of the given (x, y).

Example 1 Under a strict application of the definition we have just described, the polar equation $\theta = \alpha$ represents the half-line shown in Fig. 3.1(b), because conventionally $r \geqslant 0$. Similarly the polar curve represented by the equation $r = 1 + 2\cos\theta$ is only defined for $-2\pi/3 \leqslant \theta \leqslant 2\pi/3$ when the convention $r \geqslant 0$ is assumed, because outside this interval r takes negative values.

Example 2 The equiangular spiral, whose polar equation is

$$r = ae^{\cot\alpha}, \quad \text{where } a > 0,$$

is the curve which makes a constant angle with the radius vector. The complete curve, part of which is shown in Fig. 3.2, is only obtained if we remove the restriction $0 \leqslant \theta < 2\pi$ and allow θ to take any real value.

Fig. 3.2

Variations from the definition of polar coordinates are in some cases conventionally accepted by some writers and examination boards, and we shall adopt the following in this chapter when appropriate. Usually it is clear, from the context, which convention is being used.

(i) Negative values of r may be defined as values of r measured from the pole O in a direction opposite to the direction of r positive. Thus the coordinate pairs (c, α) and $(-c, \pi + \alpha)$ represent the same point.

In Example 1, this extension of the definition allows the polar equation $\theta = \alpha$ to include the whole line extended indefinitely either side of the pole O in Fig. 3.1(b). This extension also allows us to include the negative values of r arising in the equation $r = 1 + 2\cos\theta$ when we consider values of θ throughout the interval $0 \leqslant \theta < 2\pi$.

(ii) In some cases, particularly for curves like spirals, it is desirable to take values of θ outside the interval $0 \leqslant \theta < 2\pi$.

With these extensions of the definition, no point of the plane has a unique polar coordinate pair, but any one of its coordinate pairs defines the point uniquely.

With such variations it proves necessary, when working with polar coordinates, to consider the sketching of many curves from first principles.

3.2 Curve sketching in polar coordinates

Often the general shape of a curve whose equation is given in polar form is best investigated by finding the coordinates of a number of important points on the curve. The following techniques may also be of help and they should be employed whenever appropriate.

1. Symmetrical properties

For the curve $f(r, \theta) = 0$,

(i) symmetry about the initial line occurs when $f(r, \theta) = f(r, -\theta)$,

(ii) symmetry about the line $\theta = \pi/2$ occurs when $f(r, \theta) = f(r, \pi - \theta)$,

(iii) the curve has point symmetry about the pole O when only even powers of r occur in the equation $f(r, \theta) = 0$.

2. Aids from calculus

For the curve $r = f(\theta)$, $\qquad\qquad \dfrac{dr}{d\theta} = f'(\theta)$.

Investigate the sign of $f'(\theta)$ to find values of θ for which r increases or decreases with θ.

The angle ϕ between the tangent to the curve at a particular point $P(r, \theta)$ and the radius vector $\overrightarrow{OP}$ is given by $\tan \phi = r\dfrac{d\theta}{dr}$.

Values of θ for which $\dfrac{d}{d\theta}\left[f(\theta) \sin \theta \right] = 0$ and $\dfrac{d}{d\theta}\left[f(\theta) \cos \theta \right] = 0$ give points on the curve at which tangents are parallel to and perpendicular to the initial line, respectively.

3. Special limitations on the values of r or θ

From their equations it is often immediately clear that some curves are wholly confined to certain regions of the coordinate plane. The polar curve $r = a \sin 2\theta$, where $a > 0$, cannot lie outside the circle $r = a$, since $|\sin 2\theta| \leqslant 1$ for all θ.

4. The form of a curve at the pole

If, as $r \to 0$, θ approaches some definite angle α, then the line(s) $\theta = \alpha$ give(s) the direction(s) of the tangent(s) to the curve at the pole. Such values can often be found by solving for θ the equation $r = 0$.

5. Asymptotes and asymptotic circles

A general method of finding asymptotes and asymptotic circles to polar curves requires advanced techniques and is beyond the level discussed here. However, it is often possible to determine a limiting value for θ as $r \to \infty$ or for r as $\theta \to \infty$.

For example, the polar curve $r = a(\theta + 1)/\theta$, where $a > 0$, has an asymptotic circle $r = a$ because as $\theta \to \infty$ it is clear that $r = a(1 + 1/\theta) \to a$.

It is also possible in some cases to convert a polar equation into its cartesian form and use the methods discussed in chapter 1 concerning asymptotes.

The following examples, in all of which $a > 0$, illustrate the methods used when sketching curves from their polar equations.

Example 3   Sketch the curve $r = a \sec^2 (\theta/2)$. $[r = 2a/(1 + \cos \theta).]$

The curve is symmetrical about the initial line and the whole curve is obtained when values of θ are taken from the interval $0 \leqslant \theta < 2\pi$. However, because of the symmetry, we need only consider values of θ in the interval $0 \leqslant \theta \leqslant \pi$.

Since $\dfrac{dr}{d\theta} = a \tan (\theta/2) \sec^2 (\theta/2)$, r steadily increases as θ increases from 0 to π.

When $\theta = 0$, $r = a$.
When $\theta = \pi/2$, $r = 2a$.
As $\theta \to \pi-$, $r \to +\infty$.
A sketch of this curve, a parabola, is shown in Fig. 3.3(a). In cartesian form, the equation of the curve is $y^2 = 4a(a - x)$.

Example 4   Sketch the curve $r = a(2 \sin \theta - \operatorname{cosec} \theta)$ for $0 < \theta < \pi$.

Since $2 \sin \theta - \operatorname{cosec} \theta = 2 \sin (\pi - \theta) - \operatorname{cosec} (\pi - \theta)$, the curve is symmetrical about the line $\theta = \pi/2$.
As $\theta \to 0+$, $r \to -\infty$ and $r \sin \theta \to -a+$ from above;
as $\theta \to \pi-$, $r \to -\infty$ and $r \sin \theta \to -a+$ from above;
$\Rightarrow$ line $r \sin \theta = -a$ is an asymptote to the curve.

θ	$\pi/6$	$\pi/4$	$\pi/3$	$\pi/2$
r	$-a$	0	$a/\sqrt{3}$	a

The directions of the tangents to the curve at the pole O are given by

$$2 \sin \theta - \operatorname{cosec} \theta = 0; \text{ that is, } \theta = \pi/4 \text{ or } \theta = 3\pi/4.$$

The curve, a strophoid, is shown in Fig. 3.3(b).
The cartesian equation of the curve is $y(x^2 + y^2) + a(x^2 - y^2) = 0$.

Example 5   Sketch the rose-curve $r = 4a \sin 3\theta$ and find the area of a petal.

Since $\sin 3\theta = 3 \sin \theta - 4 \sin^3 \theta$, the curve is symmetrical about the line $\theta = \pi/2$. The curve lies within the circle $r = 4a$ and touches the circle where $\theta = \pi/6$, $5\pi/6$ and $3\pi/2$ (see chapter 2: Example 8, page 37).

Since conventionally $r \geqslant 0$, the curve does not exist in the intervals $\pi/3 < \theta < 2\pi/3$, $\pi < \theta < 4\pi/3$ and $5\pi/3 < \theta < 2\pi$ for which $r < 0$.

For $0 < \theta < \pi/6$, r increases from 0 to $4a$.

For $\pi/6 < \theta < \pi/3$, r decreases from $4a$ to 0.

The interval $0 < \theta < \pi/3$ produces a loop, symmetrical about the half-line $\theta = \pi/6$. As θ takes values in the intervals $2\pi/3 \leqslant \theta \leqslant \pi$ and $4\pi/3 \leqslant \theta \leqslant 5\pi/3$, two further identical loops are produced symmetrical about the half-lines $\theta = 5\pi/6$ and $\theta = 3\pi/2$ respectively.

The curve, a rose-curve called a *rhodonea*, is shown in Fig. 3.3(c).

The directions of the tangents to the curve at the pole O are given by $\sin 3\theta = 0$. The tangents are the half-lines $\theta = 0$, $\theta = \pi/3$, $\theta = 2\pi/3$, $\theta = \pi$, $\theta = 4\pi/3$ and $\theta = 5\pi/3$.

The points on the curve at which tangents parallel to the initial line occur are determined by finding the values of θ for which $r \sin \theta$, that is $4a \sin 3\theta \sin \theta$, has stationary values.

The area of a petal for this curve is given by

$$\frac{1}{2} \int_0^{\pi/3} r^2 \, d\theta = 8a^2 \int_0^{\pi/3} \sin^2 3\theta \, d\theta$$

$$= 4a^2 \int_0^{\pi/3} (1 - \cos 6\theta) \, d\theta$$

$$= 4a^2 \left[\theta - (\sin 6\theta)/6 \right]_0^{\pi/3} = 4\pi a^2/3.$$

Note: If the convention which allows r to take both positive and negative values had been used in this example, the same curve would be obtained by taking values of θ in the interval $0 \leqslant \theta < \pi$.

Working exercise

The family of curves called limaçons are shown in Fig. 3.3(d).

For C_1, $b < a$ $(r \geqslant 0)$.

For C_2, called a *cardioid*, $b = a$ $(r \geqslant 0)$.

For C_3, which has an inner and outer loop, $b > a$ (r may take both positive and negative values).

Analyse these equations in a similar way to that shown in the previous examples, taking suitable values of a and b.

Exercise 3.2

1 The table gives the equations of straight lines and circles in polar and equivalent cartesian forms. Verify the conversions and sketch the graphs.

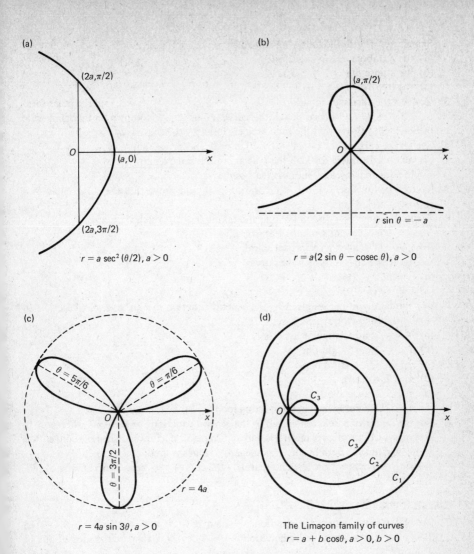

(a)

$(2a, \pi/2)$

O

$(a, 0)$

x

$(2a, 3\pi/2)$

$r = a \sec^2 (\theta/2), a > 0$

(b)

$(a, \pi/2)$

O

x

$r \sin \theta = -a$

$r = a(2 \sin \theta - \cosec \theta), a > 0$

(c)

$\theta = 5\pi/6$

$\theta = \pi/6$

O

x

$\theta = 3\pi/2$

$r = 4a$

$r = 4a \sin 3\theta, a > 0$

(d)

C_3

O

C_3

C_2

C_1

x

The Limaçon family of curves
$r = a + b \cos\theta, a > 0, b > 0$

Fig. 3.3

Description	Polar equation	Cartesian equation
Half-line through O	$\theta = \alpha$	$y = x \tan \alpha$
Straight line not through O	$r \cos (\theta - \alpha) = p$	$x \cos \alpha + y \sin \alpha = p$
Circle, centre O, radius a	$r = a$	$x^2 + y^2 = a^2$
Circle, centre (r_1, θ_1), radius R	$R^2 = r^2 + r_1{}^2 - 2rr_1 \cos (\theta - \theta_1)$	$(x - r_1 \cos \theta_1)^2 + (y - r_1 \sin \theta_1)^2 = R^2$

2   Sketch, for $a > 0$, the following named curves
    (a) the parabola $a/r = 1 + \cos\theta$,
    (b) the ellipse $a/r = 1 + \frac{1}{2}\cos\theta$,
    (c) the hyperbola $a/r = 1 + 2\cos\theta$.

3   Convert the cartesian equation $(x^2 + y^2)^2 = a^2(x^2 - y^2)$, $a > 0$, into the polar form $r^2 = a^2 \cos 2\theta$ and hence sketch the curve. (This curve is known as the lemniscate of Bernoulli.) Prove that the total area enclosed by the curve is a^2.

4   Sketch the cardioid $r = 4a(1 + \cos\theta)$, $a > 0$, and prove that
    (a) all chords drawn through the pole are of the same length,
    (b) the area enclosed by the curve is $24\pi a^2$.

5   Sketch the limaçon $r = a(5 + 4\cos\theta)$, $a > 0$, and prove that the curve encloses a region of area $33\pi a^2$.

6   Using the same pole and initial line and taking $a > 0$, sketch the curves
    (a) $r = a\theta$, called an Archimedean spiral,
    (b) $r = a\theta^{-1}$, called a reciprocal spiral.
    Use your sketches to sketch the curves
    (c) $r = a(1 + \theta)/\theta$,
    (d) $r = a\theta/(1 + \theta)$.

7   For the family of curves given by the general equation $r^n = a^n \cos n\theta$, where $a > 0$, sketch the following special cases:
    (a) $n = -2$, a rectangular hyperbola,
    (b) $n = -1$, a straight line,
    (c) $n = -1/2$, a parabola,
    (d) $n = 1$, a circle,
    (e) $n = 1/2$, a cardioid,
    (f) $n = 3/2$, a curve having intersecting loops.

8   For the family of rose-curves given by the general equation $r = a \cos n\theta$, where $a > 0$, explain why there are n petals if n is odd and $2n$ petals if n is even, when the convention allowing r to take both positive and negative values is used.
    Using this convention, sketch separately the curves $r = a \cos 3\theta$ and $r = a \cos 4\theta$.

Miscellaneous Exercise 3

1   Sketch the curve with polar equation $r = 2 \sin 2\theta$ for $0 \leqslant \theta \leqslant \pi/2$.   (L)
2   Sketch the curve whose polar equation is $r = 3 \cos 2\theta$. Calculate the area of one of its loops.   (AEB)
3   Sketch for $-\pi/4 \leqslant \theta \leqslant \pi/4$ the curve with polar equation

$$r = 4 \cos 2\theta.$$ (L)

4   Sketch the curve whose polar equation is $r = \tan \frac{1}{2}\theta$ and find the area enclosed by its loop.   (AEB)
5   Sketch the curve with polar equation $r = 2a(\sec\theta - 1)$, showing its asymptote $r \cos\theta = 2a$, where $a > 0$.   (C)
6   Sketch the curves whose polar equations are $r = a(1 + \sin\theta)$ and $r = 3a \sin\theta$, where $a > 0$.   (JMB)
7   Two curves have the equations in polar coordinates

$$r = 2 \sin\theta \quad \text{and} \quad r = 4 \sin^2\theta \quad (0 \leqslant \theta \leqslant \pi).$$

Show that the curves intersect at the pole O and at two other points A and B, giving the polar coordinates of A and B. Sketch the two curves on the same diagram.   (JMB)

8 Sketch on the same diagram the circle with polar equation $r = 4 \cos \theta$ and the line with polar equation $r = 2 \sec \theta$. State polar coordinates for their points of intersection. (L)

9 In separate diagrams sketch the curves whose equations in polar coordinates are
(a) $r = 1 + \theta$, $0 \leqslant \theta \leqslant 2\pi$,
(b) $r = 2 \sin \theta$, $0 \leqslant \theta \leqslant \pi$. (L)

10 Sketch the curve whose polar equation is $r = 2a \cos 2\theta$, where $a > 0$, and calculate the area of one of its loops. (AEB)

11 Sketch the curve whose equation in polar coordinates is $r = \sin 3\theta$.
    Find the area of the region for which

$$0 \leqslant r \leqslant \sin 3\theta, \quad 0 \leqslant \theta \leqslant \pi/3. \qquad \text{(L)}$$

12 Sketch the curve with polar equation $r = a \sin 2\theta \sin \theta$, where $a > 0$. Find the area of the region enclosed by one loop of the curve. (C)

13 Sketch the curve $r = a\theta$, where a is a positive constant and $0 \leqslant \theta \leqslant \pi$, and find the area of the region enclosed by this curve and the half-line $\theta = \pi$. (L)

14 Sketch the curve with equation

$$r = a \cos 2\theta, \quad \text{where } a > 0, \quad \text{for } -\pi/4 \leqslant \theta \leqslant \pi/4,$$

and find the area of the region for which $0 \leqslant r \leqslant a \cos 2\theta$, $-\pi/4 \leqslant \theta \leqslant \pi/4$.
    By considering the stationary values of $\sin \theta \cos 2\theta$, or otherwise, find the polar coordinates of the points on the curve where the tangent is parallel to the initial line. (L)

15 Sketch the curves whose equations in polar coordinates are
(i) $r = a(1 + \cos \theta)$ $(a > 0)$,
(ii) $r^2 = 4 \cos 3\theta$ $(r > 0)$.
    Find the total length of arc of the curve corresponding to (i).
    Find also the area of one of the loops of the curve corresponding to (ii). (C)

16 Sketch the curve whose polar equation is $r = 2(1 + \cos \theta)$. Show that this curve and the curve whose polar equation is $r(1 + \cos \theta) = 2$ cut at right angles at the points $(2, \pm\frac{1}{2}\pi)$. Sketch the region defined by the inequalities $r \leqslant 2(1 + \cos \theta)$, $r \leqslant 2/(1 + \cos \theta)$, and calculate its area. (JMB)

17 Shade the region C for which the polar coordinates r, θ satisfy

$$r \leqslant 4 \cos 2\theta \quad \text{for } -\pi/4 \leqslant \theta \leqslant \pi/4.$$

Find the area of C.
    State the equations of the tangents to the curve $r = 4 \cos 2\theta$ at the pole. (L)

18 Show that the curve K whose polar equation is

$$r = -2 \cos \theta + 2\sqrt{(\cos^2 \theta + 3)}$$

is a circle whose centre A has polar coordinates $(2, \pi)$ and that it touches the curve C whose polar equation is

$$r = 3(1 - \cos \theta)$$

at the point B where $\theta = \pi$. Sketch the curves C and K on the same diagram. (JMB)

19 Sketch the curve whose polar equation is $r^2 = 2 \sin 2\theta$. Find the area enclosed by a loop of the curve.
    Find the cartesian equation of the curve and hence show that $x = \dfrac{2t}{1 + t^4}, y = \dfrac{2t^3}{1 + t^4}$
    may be taken as parametric equations of the curve.

Find the values of t for the points on the curve at which the tangents are parallel to the y-axis.

(AEB)

20 Sketch the curve with polar equation $r = a \cos \theta(2 - \cos \theta)$, where $a > 0$.

Prove that the area of the region enclosed between the inner and outer loops of the curve is $\frac{16}{3}a^2$.

(C)

Answers

Exercise 1.3

2 $2x - y - 3 = 0$

4 (a) $\tan^{-1}(7/4)$, (b) $126/13$

5 $(-2, 4), (-3, 1); (4, 2), (3, -1)$

6 $(1, 2)$.

Exercise 1.5

1 (a) (i) $x + 2 = 0$, (ii) $y \geqslant 0$;
(b) (i) $x - 2 = 0$, (ii) $y \geqslant -8$;
(c) (i) $x = 0$, (ii) $y \geqslant 4$;
(d) (i) $x = 0$, (ii) $y \leqslant 4$;
(e) (i) $2x - 1 = 0$, (ii) $y \leqslant -15/4$

5 $p = 4/9, q = -3, r = 6$.

Exercise 1.6

3 (a) even, (b) odd, (c) none of these, (d) periodic, period 3

4 $f(x) = x$ for $0 \leqslant x < 1$,
$f(x) = 4 - 2x$ for $1 \leqslant x < 3$,
$f(x) = f(x + 3)$ for $x \in \mathbb{R}$

5 (a) odd, (b) even, (c) even.

Exercise 1.9

7 (a) $y - x = 0$, (b) $y = 0$,
(c) $y = 0$

8 (a) $x = 0, y = x - 3$, (b) $x = y$,
$y = 2^{1/2}, y = -2^{1/2}$, (c) $y = x + 1$,
$y = -x + 1$

9 (a) $y = (x + 1)^2$,
(b) $4x^2 + y^2 = 1$, (c) $x^3 = (x + y)^2$,
(d) $(x^2 + y^2)^2 = 4xy$

Miscellaneous Exercise 1

1 $(8,2)$

7 $y = 2/3, x = 4/3, x = -1$

8 $\{x: 5/3 < x < 5\}$

9 $(-1, 0)$

10 $(1, 0), (-1, 6), (3, 2)$

(i) $(3/2, -1/4)$, (ii) $x = 0, y = 3$,

11 $x = 0, x = -2$

12 k/x

15 $(0, -1/4); x = 2, x = -2, y = 1$

16 $(-1, 0); 1, -3; (\frac{1}{2})^{1/3}$, min.

17 $\{y: -5/4 \leqslant y \leqslant 5\}; 5/2$

18 $2 \pm \sqrt{3}, 2 \pm \sqrt{3}; x + 1 = 0$,
$x + 3 = 0, y = 0$

20 $x \in \mathbb{R}, x \neq 3, x \neq -1; y \geqslant \frac{1}{2}$,
$y \leqslant -1$

21 $x - 4ty + 2t^2 = 0$;
$x + 4y + 2 = 0, x - 4y + 2 = 0$

24 $\{y \in \mathbb{R}, y \geqslant 3\}$

25 $y \in \mathbb{R}, y \neq 2$;
$g^{-1}: x \mapsto (x + 1)/(x - 2)$, where
$x \in \mathbb{R}, x \neq 2$

26 (iii) $3(t^2 + 1)/(2t)$

27 (ii) $m(x) \geqslant 0$

28 $x = 0, x = 4, y = 2$; max. $(1, -1)$;
min. $(-2, 5/4)$

29 $x = 2/3, y$ is a min.; $x = 2, y$ is a max.

32 (i) $x - 2 = 0, y - 1 = 0$;
(ii) $x + 2y = 0$

33 $1/3 \leqslant \lambda \leqslant 3$.

36 (b) odd, neither, even

37 $2x + 8(1 - x)^{-2}, 2 + 16(1 - x)^{-3}$,
min., $(3, 5)$

38 (i) $(0, 2), (2, -4)$; (ii) $(-2, 0)$,
$(4, -2)$

40 $4a^3 + 2ab + a^2 y' + 2by' = 0$,
$12a^2 + 2b + 4ay' + a^2 y'' + 2y'^2 +$
$2by'' = 0$. $(0, \sqrt{3})$ and $(0, -\sqrt{3})$ max.;
$(1, -2)$ and $(-1, -2)$ min.

Exercise 2.3

11 (a) $\pi/4$, (b) $-\pi/6$, (c) π,
(d) $-\pi/6$, (e) $\pi/6$, (f) $-\pi/6$.

Miscellaneous Exercise 2

3 1 root; $x + y - 2 = 0$

5 $\pi/6, \pi/2, 5\pi/6$

8 min., max., min.; $\pm 0.79, \pm 1.32$

9 $\pi/6, 5\pi/6, 3\pi/2; y = 3\sqrt{3}/2$ (max),
$y = -3\sqrt{3}/2$ (min.), $y = 0$ (neither)

11 g: (a) 2π, (b) even; h: (a) 3π,
(b) neither

12 $(1 - \sin\theta - 2\sin^2\theta)/\cos\theta$;
$(1/2, 3\sqrt{3}/4), (1/2, -3\sqrt{3}/4), (1, 0)$

13 $-1/e$

16 25, 16·26°; 25, −25, 180n° + 73·74°

17 −x + 2x^3/3; x + y = 0

18 −2(1 − x^2)$^{1/2}$

20 Max.; k = e^{-1}

21 $\sqrt{52}$ cos (x + 56·3°);
(123·7°, −$\sqrt{52}$); 33·7°, 213·7°

23 π/6, 5π/6, −π/2, π/2;
−π ⩽ θ ⩽ −π/2, π/6 ⩽ θ ⩽ π/2,
5π/6 ⩽ θ ⩽ π

25 (i) f(x) ⩾ 1 + ln 2,
(ii) f^{-1}(x) = (e^{x-1} − 2)/3,
(iii) x ⩾ 1 + ln 2

27 (1, e^{-1}); 0, n : n even, min. (0, 0),
max. (n, n^ne^{-n}); n odd, inflexion (0, 0),
max. (n, n^ne^{-n})

29 π/4

30 −(b tan θ)/a

Miscellaneous Exercise 3

2 9π/8

4 2 − π/2

7 (1, 5π/6), (1, π/6)

8 (2$\sqrt{2}$, π/4), (2$\sqrt{2}$, 7π/4)

10 πa^2/2

11 π/12

12 πa^2/16

13 $\pi^3 a^2$/6

14 πa^2/8, (2a/3, ±0·42)

15 8a; 4/3

16 (9π − 16)/3

17 2π; θ = π/4, θ = −π/4

19 1; (x^2 + y^2)^2 = 4xy; ±3$^{-1/4}$, ∞

Index